AF864339

Springer Theses

Recognizing Outstanding Ph.D. Research

Aims and Scope

The series "Springer Theses" brings together a selection of the very best Ph.D. theses from around the world and across the physical sciences. Nominated and endorsed by two recognized specialists, each published volume has been selected for its scientific excellence and the high impact of its contents for the pertinent field of research. For greater accessibility to non-specialists, the published versions include an extended introduction, as well as a foreword by the student's supervisor explaining the special relevance of the work for the field. As a whole, the series will provide a valuable resource both for newcomers to the research fields described, and for other scientists seeking detailed background information on special questions. Finally, it provides an accredited documentation of the valuable contributions made by today's younger generation of scientists.

Theses may be nominated for publication in this series by heads of department at internationally leading universities or institutes and should fulfill all of the following criteria

- They must be written in good English.
- The topic should fall within the confines of Chemistry, Physics, Earth Sciences, Engineering and related interdisciplinary fields such as Materials, Nanoscience, Chemical Engineering, Complex Systems and Biophysics.
- The work reported in the thesis must represent a significant scientific advance.
- If the thesis includes previously published material, permission to reproduce this must be gained from the respective copyright holder (a maximum 30% of the thesis should be a verbatim reproduction from the author's previous publications).
- They must have been examined and passed during the 12 months prior to nomination.
- Each thesis should include a foreword by the supervisor outlining the significance of its content.
- The theses should have a clearly defined structure including an introduction accessible to new PhD students and scientists not expert in the relevant field.

Indexed by zbMATH.

More information about this series at http://www.springer.com/series/8790

Akihisa Miyagawa

Acoustic Levitation-Based Trace-Level Biosensing

Design of Detection Systems and Applications to Real Samples

Doctoral Thesis accepted by
the Tokyo Institute of Technology, Tokyo, Japan

Springer

Author
Dr. Akihisa Miyagawa
University of Tsukuba
Tsukuba, Ibaraki, Japan

Department of Chemistry
School of Science
Tokyo Institute of Technology,
Tokyo, Japan

Supervisor
Prof. Tetsuo Okada
Tokyo Institute of Technology
Tokyo, Japan

ISSN 2190-5053 ISSN 2190-5061 (electronic)
Springer Theses
ISBN 978-981-16-1424-8 ISBN 978-981-16-1425-5 (eBook)
https://doi.org/10.1007/978-981-16-1425-5

© The Editor(s) (if applicable) and The Author(s), under exclusive license to Springer Nature Singapore Pte Ltd. 2021
This work is subject to copyright. All rights are solely and exclusively licensed by the Publisher, whether the whole or part of the material is concerned, specifically the rights of translation, reprinting, reuse of illustrations, recitation, broadcasting, reproduction on microfilms or in any other physical way, and transmission or information storage and retrieval, electronic adaptation, computer software, or by similar or dissimilar methodology now known or hereafter developed.
The use of general descriptive names, registered names, trademarks, service marks, etc. in this publication does not imply, even in the absence of a specific statement, that such names are exempt from the relevant protective laws and regulations and therefore free for general use.
The publisher, the authors and the editors are safe to assume that the advice and information in this book are believed to be true and accurate at the date of publication. Neither the publisher nor the authors or the editors give a warranty, expressed or implied, with respect to the material contained herein or for any errors or omissions that may have been made. The publisher remains neutral with regard to jurisdictional claims in published maps and institutional affiliations.

This Springer imprint is published by the registered company Springer Nature Singapore Pte Ltd.
The registered company address is: 152 Beach Road, #21-01/04 Gateway East, Singapore 189721, Singapore

Supervisor's Foreword

Nano/microparticles provide versatile ways for designing reactions and analyses that could not be put into practice without their utilization. Particle manipulation is one of the critical tasks for the successful development of the systems, in which particles take an essential part. An external physical force has been used as a powerful tool for this purpose. Because physical forces are mostly functions of particle sizes, particles with different sizes behave differently. This intrinsic nature of physical forces gives a principle for particle separation in terms of size but hinders the detection of a difference in its properties other than size. We developed a combined acoustic-gravitational (CAG) field, which recognizes the acoustic properties (density and compressibility) of microparticles but does not detect their sizes. The CAG field was applied to the separation of particles and recognition of ions bound to an ion-exchange resin bead.

Dr. Miyagawa's doctoral thesis describes particle-based bioanalytical methods, in which he proposes a novel approach allowing the transduction of the density of a polymer microparticle into its levitation coordinate in the CAG field. The particle density is modified by the binding of gold nanoparticles on the polymer microparticle. This approach offers very high sensitivity and versatility. Sensitivity arises from the small dimension of the microparticle; the binding of a small number of gold nanoparticles (typically several hundred) causes a detectable change in the levitation coordinate of the microparticle. This density change can be induced by various reactions, avidin–biotin reaction, nucleic acid hybridization, aptamer–target complexation, etc. Thus, this principle is readily applicable to various systems and allows highly sensitive (typically pM or lower) detection of target compounds of medical, physiological, and environmental importance. I believe that Dr. Miyagawa's work opens a new door in bioanalysis.

Tokyo, Japan
January 2021

Prof. Tetsuo Okada

Parts of this thesis have been published in the following journal articles:

1. A. Miyagawa, and T. Okada, Particle Manipulation with External Field; from Recent Advancement to Perspectives., *Anal. Sci.*, 2021, **37**, 69–78.
2. A. Miyagawa, Y. Okada and T. Okada, Aptamer-Based Sensing of Small Organic Molecules by Measuring Levitation Coordinate of Single Microsphere in combined Acoustic-Gravitational Field, *ACS Omega 2020, 5*, 3542–3549.
3. A. Miyagawa, and T. Okada, Trace Reaction Measurements Using Acoustic Levitation of a Single Particle, *Bunseki Kagaku 2019, 68*, 549–558.
4. A. Miyagawa, M. Harada and T. Okada, Multiple MicroRNA Quantification Based on Acoustic Levitation of Single Microspheres after One-Pot Sandwich Interparticle Hybridizations, *Anal. Chem. 2018, 90*, 13729–13735.
5. A. Miyagawa, M. Harada and T. Okada, Zeptomole Biosensing of DNA with Flexible Selectivity Based on Acoustic Levitation of a Single Microsphere Binding Gold Nanoparticles by Hybridization, *ACS Sens. 2018, 3*, 1870–1875.
6. A. Miyagawa, M. Harada and T. Okada, Zeptomole Detection Scheme Based on Levitation Coordinate Measurements of a Single Microparticle in a Coupled Acoustic-Gravitational Field, *Anal. Chem. 2018, 90*, 2310–2316.
7. A. Miyagawa, M. Harada and T. Okada, Detection of Microsphere Surface Reaction Using Acoustic-Gravity Field., *Proc. Acoustofluidics 2017 2017*, 66.
8. A. Miyagawa, Y. Inoue, M. Harada and T. Okada, Acoustic Sensing Based on Density Shift of Microspheres by Surface Binding of Gold nanoparticles, *Anal. Sci. 2017, 33*, 939-944.

Acknowledgements

First, I would like to express my sincere gratitude to Prof. Tetsuo Okada, Department of Chemistry at Tokyo Institute of Technology, for the constructive discussion and excellent advices about my studies. Thanks to his unique ideas, discussions, broad knowledge, and support, I have considerably evolved during my Master's and Ph.D. courses with respect to writing skills in Japanese and English, ideas, and knowledge. I also thank him for adopting my new themes based on various concepts and letting me experiment. I would like to acknowledge Associate Professor Gaku Fukuhara, Department of Chemistry at Tokyo Institute of Technology, for several stimulating discussions and valuable career advice. I thank him for adopting my idea and new concept using his high-pressure apparatus and advising about the experiments. I thank Assistant Professor Makoto Harada, Department of Chemistry at Tokyo Institute of Technology, for several discussions and experimental support for the calculations using MOPAC. I also thank Assistant Professor Takuhiro Otsuka, Department of Chemistry at the Tokyo Institute of Technology, for several discussions and experimental support. Their ideas and concepts will remarkably influence my career as a scientist.

I am grateful to the former supervisor, Prof. Akihide Hibara, Institute of Multidisciplinary Research for Advanced Materials at Tohoku University, for his valuable comments and excellent discussions. I also thank Dr. Yoshinori, Inoue, Department of Applied Chemistry at Aichi Institute of Technology, for teaching the preparation of epoxy particles and the discussions. I thank Drs. Mizuo Maeda and Tohru Takarada, RIKEN for helpful discussions and valuable advice. I also thank Assistant Professor Arinori Inagawa, Department of Applied Chemistry at Utsunomiya University, for excellent discussions and helpful career advice. I am grateful to the secretary in Okada laboratory, Ms. Naoko Takagi, for office processing for business trips and orders of reagents and supplies. I also thank Mr. Yusuke Okada, my junior in Okada laboratory, for teaching and helping the calculation using MatLab. I am also grateful to Ms. Yue Zheng at Thinghua University for helping me with my study and stimulating discussions.

I am profoundly thankful for all discussions with former and present members in Okada, Fukuhara, and Hibara laboratories, Mr. Yusuke Iimura, Mr. Masaya Shimizu, Mr. Kouki Tokumasu, Mr. Yuta Nomura, Mr. Kotohiro Furukawa, Mr.

Kyohei Ishikawa, Ms. Aoi Akiyama, Mr. Kensuke Yanagisawa, Mr. Yoshiharu Fukui, Mr. Takuya Endo, Ms. Lin Zhou, Ms. Saori Fujino, Ms. Mana Shimohira, Mr. Koki Iijima, Ms. Hinako Sakai, Mr. Tomoaki Tsuchiya, Mr. Yu Fukunaga, Mr. Tomoya Muto, Ms. Minori Doi, Mr. Hiroaki Mizuno, Mr. Sho Suzuki, Ms. Kotoe Nakasha, Ms. Yuma Ryoson, Ms. Minami Fukuchi, Mr. Yuga Yashima, Mr. Shun Kataoka, Mr. Reiya Nishi, and Mr. Tomokazu Kinoshita. They also provided a stimulating and fun environment and enriched my time at the Tokyo Institute of Technology.

I express my appreciation toward the Grant-in-Aids for the Sasakawa Scientific Research Grant by the Japan Science Society and the Japan Society of Promotion of Science Fellows. My work was supported by these grants. I also thank the scholarship provided by the Japan Student Services Organization for financial support.

I am grateful to Mr. Katsuaki, Hori, Mr. Ryouhei, Kikuchi, Mr. Jun, Koki, Mr. Yoshihiro Kanemaki, and Mr. Masaru Tada, Ookayama Materials Division, Technical Department at Tokyo Institute of Technology, for teaching and helping with the SEM, FE-SEM, and TEM observations.

Finally, I would like to thank my father, Tsutomu Miyagawa, my mother, Youko Miyagawa, my older brother, Shunichi Miyagawa, and my younger brother, Tomohiro Miyagawa, for supporting me for a long time. Without their encouragement, it would have been impossible for me to finish this work.

Contents

Chapter 1
Introduction

Abstract In this chapter, the background of ultrasound and the application of particle separation using various external fields are introduced. The features and applications of ultrasound are discussed in Sect. 1.1. Particle separation and manipulation using external fields, such as dielectric, magnetic, electric, optical, thermal, and acoustic fields, are reviewed in Sects. 1.2 and 1.3. Finally, the purpose of this thesis is described in Sect. 1.4.

Keywords Ultrasound · External fields · Separation · Manipulation

1.1 Background

A sound wave with a frequency higher than 20 kHz is classified as ultrasound, which is used in a wide variety of fields such as environmental science, mechanical science, biochemistry, food science, and cancer therapy [1–4]. Animals such as bats and dolphins use ultrasound to obtain various information, for example, the distance, direction, and size of an object, which is called echolocation. Thus, ultrasound is omnipresent around us. A piezoelectric element is usually used for the generation of ultrasound. Ultrasound applications are typically classified into two categories: non-invasive analysis, based on the reflection of the ultrasound and mechanical uses of its destructive force. Information obtained from the reflected ultrasound enables the measurement of the distance from the liquid surface to the depth of the sea, the detection of cracks in a solid substance, and the acquisition of diagnostic images, including fetal imaging. Bioimaging using ultrasound contrast agents such as nano/microbubbles and photoacoustic reagents has rapidly developed in the last 20 years [5–8].

Ultrasound-induced energy is also useful for washing, atomization, welding, and cutting. For example, chemists often use an ultrasound washing machine for emulsification, [9, 10] cleaning, [11] dispersion, [12, 13] and dissolving a solute [14]. The radiation of the ultrasound to a solution generates bubbles, which are stretched and collapsed by pressure fluctuations. The bubbles collapse causes cavitation, which has been used for various purposes. Ultrasonic cavitation induces local high temperature and pressure and facilitates chemical reactions [15–17]. When a bubble collapses,

© The Author(s), under exclusive license to Springer Nature Singapore Pte Ltd. 2021

A. Miyagawa, *Acoustic Levitation-Based Trace-Level Biosensing*, Springer Theses,
https://doi.org/10.1007/978-981-16-1425-5_1

the temperature and pressure reach 5000 K and 1000 atm, respectively. This field of locally high temperature and pressure is called a hot spot, [18] which is used for inorganic or organic synthesis, [19–21] degradation of harmful matters, [22, 23] and cutting polymer chains [24]. Therefore, ultrasound cavitation has attracted considerable attention in various fields such as organic chemistry, physical chemistry, material, and environmental sciences [15–17].

An acoustic radiation force is exerted on a substance with an acoustic impedance that is different from that of the medium, for example, a solid material in solution [25–27]. The acoustic radiation force has been mainly used for particle separation and manipulation, [28–31] which are important in bioanalysis, chemical analysis, diagnostics, and environmental assessment. Particle separation is classified into two techniques: passive and active modes [32, 33]. The interactions between particles, particles and microchannel structure, and particles and flow are used for the passive modes, whereas an external field such as dielectric, magnetic, optical, and acoustic fields is applied to create different particle distributions in the active modes. The separation efficiency for the active mode is usually higher than that for the passive mode. Among the active modes, the acoustic field has attracted considerable attention because of its non-contact and non-invasive nature [34].

1.2 Particle Separation and Manipulation Using Various External Fields

In this section, particle separation and manipulation techniques, using external fields, except for the acoustic field, are reviewed.

Dielectrophoresis

Dielectrophoresis (DEP) was first recognized and explored by Pohl for the separation and manipulation of particles [35]. A polarizable particle in an ununiform electric field experiences an attractive or repulsive force. The time-averaged DEP force exerted on a homogeneous spherical particle is given by [32, 33, 35–37]:

$$\langle F_{\mathrm{DEP}} \rangle = 2\pi r^3 \varepsilon_{\mathrm{m}} \mathrm{Re}[f_{\mathrm{CM}}(\omega)] \nabla \mid \underset{\mathrm{rms}}{\vec{E}} \mid^2 \quad (1.1)$$

where r is the radius of the particle, ε_{m} is the permittivity of the medium, ∇ is the gradient operator, and E_{rms} is the root mean square of the electric field. The real component of the Clausius–Mossotti (CM) factor, $\mathrm{Re}[f_{\mathrm{CM}}(\omega)]$, is followed by:

$$\mathrm{Re}[f_{\mathrm{CM}}(\omega)] = \frac{\varepsilon_{\mathrm{p}}^* - \varepsilon_{\mathrm{m}}^*}{\varepsilon_{\mathrm{p}}^* + 2\varepsilon_{\mathrm{m}}^*} \quad (1.2)$$

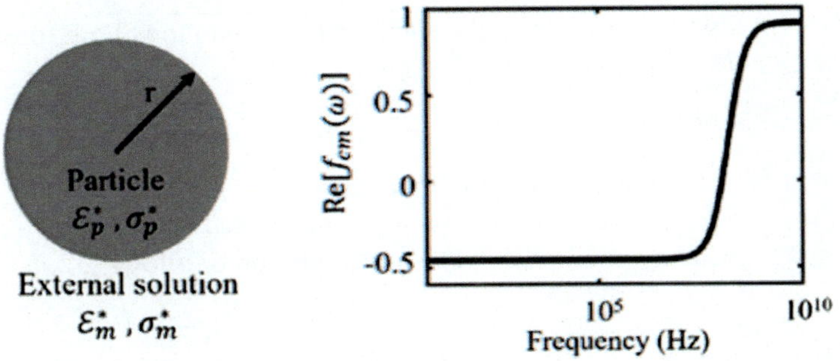

Fig. 1.1 Schematic of a homogeneous sphere particle and frequency dependence on $Re[f_{CM}(\omega)]$. The calculation was performed in the case of $\varepsilon_p{}^* < \varepsilon_m{}^*$. Reproduced from Ref. 36. Copyright

where $\varepsilon_p{}^*$ and $\varepsilon_m{}^*$ represent the complex permittivities of the particle and medium, respectively. Although $\langle F_{DEP} \rangle$ for other entities such as single-shell and multi-shell spheres and non-spheres are described by different models, [36] we focus here on the behavior of spherical homogeneous particles.

The moving direction of a particle is determined by the CM factor. When $Re[f_{CM}(\omega)] > 0$, the particle is transported toward a high electric field density (positive DEP, pDEP). In contrast, the particle moves toward a low electric field density (negative DEP, nDEP) at $Re[f_{CM}(\omega)] < 0$. Figure 1.1 shows the relationship between the frequency of an electric field and $Re[f_{CM}(\omega)]$, in which the calculation was performed in the case of $\varepsilon_p{}^* < \varepsilon_m{}^*$. With increasing frequency, $Re[f_{CM}(\omega)]$ changes from negative to positive values. This indicates that adjusting the frequency of the ununiform electric field allows the separation of particles with different $\varepsilon_p{}^*$ values.

Yunus et al. designed a microfluidic device to separate colloidal latex spheres using DEP by integrating two interdigitated electrodes on the top and bottom walls of a channel [38]. In this device, the nDEP force acted on 1 μm latex spheres, whereas no force was exerted on 500 nm latex spheres at 30 V_{pp} and 5 MHz. Thus, the sub-μm particle separation was successful in the continuous flow. Recently, Jones et al. reported continuous size-based separation using DEP [39]. Their microdevice was designed such that an ununiform electric field was generated near the corners of a bottleneck between the flow reservoir and multiple outlets. As a result, pDEP and nDEP forces were exerted on DNA molecules depending on their lengths, and they were carried to different outlet channels. Han et al. continuously separated red and white blood cells from diluted blood samples using an interdigitated electrode [40]. This electrode was placed at an appropriate angle relative to a fluid flow such that red and white blood cells experienced different DEP forces. The collecting efficiencies were 87 and 92% for the red and white blood cells, respectively. Zhao et al. recently developed an alternating current (AC)–DEP microfluidic chip for the separation of yeast cells [41]. A non-uniform electric field was generated by the electrodes embedded in asymmetric orifices on the opposite sidewalls. Living cells experienced the pDEP force from the large electric field gradient near the orifice, whereas dead cells were repelled away from the small orifice due to the nDEP force. These results were supported by numerical calculations.

Magnetic field

A magnetic field can be simply generated using a permanent magnet or electromagnet. An attractive or repulsive force is exerted on a magnetic particle in an inhomogeneous magnetic field. This magnetic force follows [32, 33, 42–44]:

$$F_{\mathrm{mag}} = \frac{4\pi r^3}{3}(\chi_{\mathrm{m}} - \chi_{\mathrm{p}})\frac{(B\nabla)B}{\mu_0} \tag{1.3}$$

where χ_{p} and χ_{m} are the magnetic susceptibilities of the particle and medium, respectively, B is the magnetic field intensity, and μ_0 is the magnetic permittivity of vacuum. Typically, a magnetic particle experiences F_{mag} of ~ pN with a small permanent magnet [45].

Zhu et al. presented a separation system for non-magnetic particles in ferrofluids [46]. The non-magnetic particles inside ferrofluids experienced magnetic and hydrodynamic drag forces. Using these forces, the size-based separation (i.e., 3.1 and 9.9 μm) of the non-magnetic particles was demonstrated. Pamme et al. developed a continuous magnetophoretic chip including multiple bifurcations [47]. A particle mixture was introduced into a chamber, in which a magnetic field perpendicular to the flow direction was generated. Magnetic particles were separated according to their sizes in this chamber because particles with larger diameters experience a larger magnetic force. Adams et al. developed a multi-target cell sorter using the magnetic field [48]. Two different target cells were labeled with magnetic beads of different diameters. The target cells were separated from non-target cells because they experience different magnetic forces. Furlani et al. reported direct and continuous separation of red and white blood cells in plasma [49]. White blood cells were regarded as diamagnetic particles, whereas red blood cells behaved as diamagnetic or paramagnetic particles depending on the oxygenation state.

A magnetic field was also used for manipulation known as magnetic tweezers [50–52]. As shown in Fig. 1.2, basic magnetic tweezers consist of a permanent magnet placed above the sample holder set on an inverted microscope. The force of magnetic tweezers typically ranges from 10^{-3} to 10^{2} pN for 0.5–5.0 μm particles [51]. The mechanochemical studies of DNA gyrase [53] and the rotary motor F0F1ATPase [54] were performed using this technique. Liepfert et al. demonstrated that double-stranded RNA (dsRNA) is similar to double-stranded DNA (dsDNA) in terms of the bending and twist persistence lengths and the force-torque phase diagram, whereas the stretch modulus of dsRNA was one-third of that of dsDNA [55] In this study, DNA and RNA, which were bound on the substrate and particle, respectively, were stretched and bent by F_{mg}. Harada et al. visualized RNA polymerase transcribing DNA using direct and real-time optical microscopy combined with magnetic tweezers [56] in which RNA polymerase was bound to the cell and a DNA molecule was anchored on a magnetic bead. They observed the in situ rotation

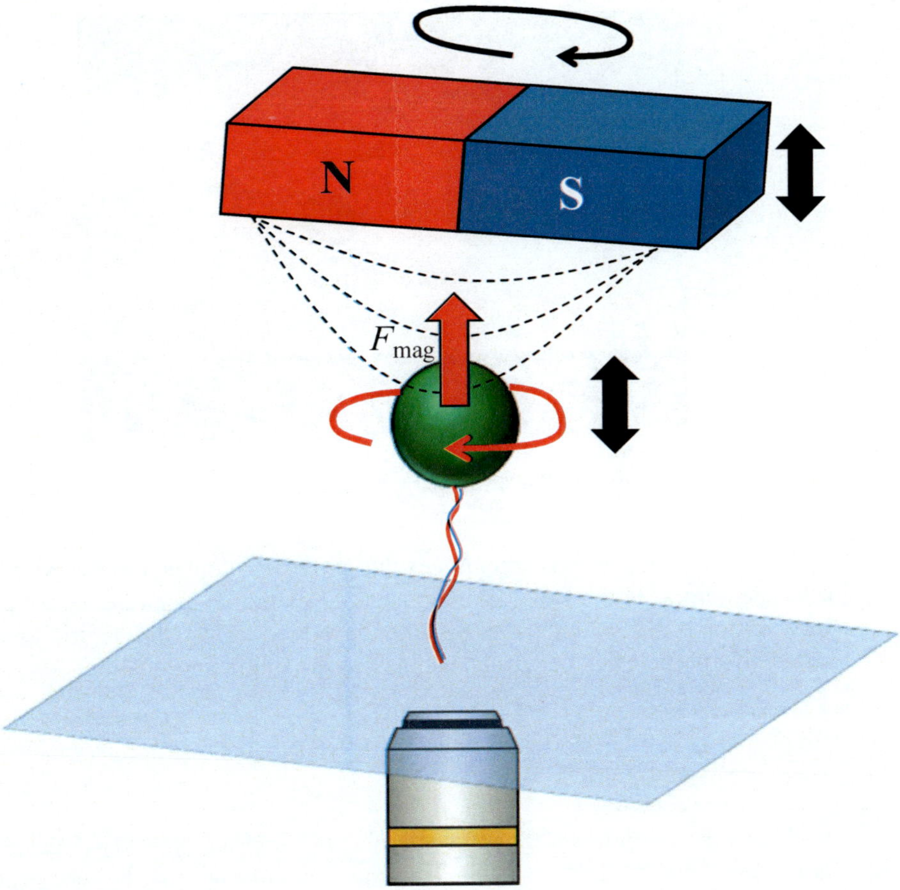

Fig. 1.2 Schematic of magnetic tweezers

of DNA when enzymatic transcription occurred. Whitesides et al. developed a novel magnetic manipulation system, called magnetic levitation (MagLev) [57–61]. Particles or liquid droplets with different densities are levitated at different coordinates due to the density recognition ability of Maglev. At the levitation coordinate, the magnetic and sedimentation forces exerted on the particles are balanced.

Electric field

The electric force is proportional to the electric field strength and particle charge. The effective charge of a particle is determined by its zeta potential and Debye length ($\lambda_D = 1/\kappa$). The Debye–Hückel parameter (κ) is expressed by:

$$\kappa = \sqrt{\frac{e^2 N_{av}}{\varepsilon k T} \sum_i \left(c_i {z_i}^2\right)} \quad (1.4)$$

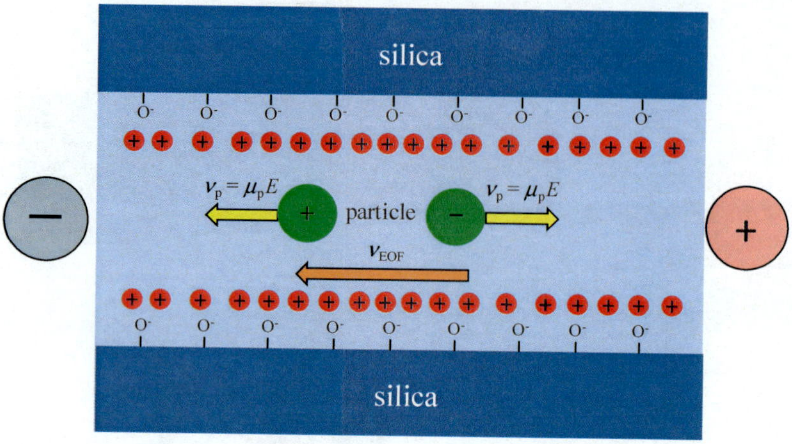

Fig. 1.3 Schematic of the CE principle

where e is the elementary charge, N_{av} is the Avogadro number, ε is the permittivity of the medium, k is the Boltzmann constant, T is the temperature, c_i is the ionic concentration of the solution, and z_i is the ion valency. Thus, the electric force (F_e) exerted on a particle can be written using K as the following equation:

$$F_e = qE = 2\pi d_p\left(1 + \frac{1}{2}\kappa d_p\right)\varepsilon\zeta E \tag{1.5}$$

where q is the particle charge, E is the electric field strength, d_p is the particle diameter, and ζ is the zeta potential. This equation indicates that F_e depends on the particle diameter and zeta potential.

The electric field is often used for electrophoretic separation. Capillary electrophoresis (CE) is an analytical technique widely applied to the separation of ions, proteins, and particles [62–64]. Figure 1.3 represents the principle of CE. When charged particles are introduced to an electric field applied along the capillary axis, they migrate to an electrode at the electrophoretic velocity (V_p) according to their electrophoretic mobility (μ_p):

$$v_p = \mu_p E \tag{1.6}$$

An electric double layer (EDL) is formed near the wall of the silica capillary because of the dissociation of the silanol groups on the silica surface. When an electric field is applied, the cations in the EDL migrate toward the cathode. This migration causes a fluid flow in the same direction, called electroosmotic flow (v_{EOF}). Thus, the apparent velocity of the particle is given by:

$$v_{app} = v_p + v_{EOF} \tag{1.7}$$

Because $\nu_{EOF} > \nu_p$ in many cases, particles flow out of the cathodic end irrespective of the sign of charge and charge density of the particle.

Differentiation of virus serotypes was performed with capillary zone electrophoresis by Okun et al. [65]. Four different human rhinoviruses (HRV2, HRV14, HRV16, and HRV49) were separated with borate buffer at pH 8.3 as the running solution. Sonohara et al. measured the μ_p of *Escherichia coli* and *Staphylococcus aureus* by changing the pH and ionic strength of the running buffer [66] They found that *E. coli* was more negatively charged than *S. aureus* because *E. coli* has a lipopolysaccharide, whereas *S. aureus* is covered with a peptidoglycan. Tsukagoshi et al. evaluated the migration behavior of polymer particles using buffers containing metal ions or saccharides [67]. The mobility of the particles decreased in the presence of multi-charged counterions because they were strongly bound to the particle surface. Armstrong et al. demonstrated high-efficiency separation of three bacteria (Pseudomonas fluorescens, Enterobacter aerogenes, and Micrococcus luteus) via the size and shape by adding polyethylene oxide (PEO) to the running buffer [68]. The migration time was controlled by varying the concentration of PEO.

Cohen et al. demonstrated a single nanoscale object trapping in solution using the electrokinetic method [69]. Four microelectrodes were placed on the glass slide such that their tips were directed to the center of the glass slide. Electrophoresis caused entrapping of a 200 nm particle at the center of the glass slide. Yazbeck et al. characterized and manipulated single nanoparticle using nanopore-based electrokinetic tweezers [70]. When a voltage bias was applied across a nanopore, a charged nanoparticle experienced F_e, F_{EOF}, and F_{DEP}. The nanoparticle was entrapped at the equilibrated position, where the three forces were balanced.

Optical field

When the light passes through a substance, there is a difference in light momentum between the incident and refracted light. The total amount of momentum should be conserved. The optical force (F_{opt}) exerted on the substance is represented by the following equation [71, 72]:

$$F_{opt} = \frac{2n_1 P}{c}\left(\frac{r}{\omega}\right)^2 Q^* \tag{1.8}$$

where n_1 is the refractive index of the medium, P is the laser power, c is the speed of the light, and Q^* is the trapping efficiency that depends on the size, shape, material of the particle, and its position with respect to the spatial profile of the light. Typically, the optical force exerted on a 0.25–5.00 μm spherical particle ranges from 0.1 to 100 pN [51, 71].

MacDonald et al. designed a microfluidic system integrating an optical lattice for particle separation, in which the amplitude and phase were different for each beam spot [73]. When particles with different properties reached an optical lattice, they were deflected from their original path according to their optical potential energies. Optical chromatography was developed by Imasaka et al. [74]. When flowing particles are illuminated by the laser beam introduced into the flow channel in the direction opposite to that of the flow, they experienced an optical radiation pressure force. Thus, depending on the size and the refractive index of the particles, they were pushed toward the focal region of the beam and then reached equilibrium positions, at which optical and flow forces are balanced. Using this method, Hebert et al. recently evaluated *Bacillus anthracis* spore uptake in macrophage cells [75].

Optical tweezers, established by Ashkin, [76] enable the trapping and manipulation of particles and cells. He showed that the difference in the refractive index between a particle and medium causes an attraction force to the center of the laser beam. Mammen et al. applied this technique to evaluate the binding probability of virus-coated beads with erythrocytes in the presence of inhibitors [77]. Abbondanzieri et al. used dual optical tweezers, which individually trapped a DNA-anchored bead and RNA polymerase-anchored bead, to investigate the mechanism of RNA polymerase translocation [78]. As shown in Fig. 1.4, as one bead is strongly entrapped while the other is weakly entrapped, the polymerase translocation starts at the location of the weakly trapped particle. When the polymerase translocation occurs, it transfers to the next DNA base pair (bp), leading to the dynamic movement of the bead because it is pulled by optical force. Discrete steps averaging 3.7 ± 0.6 Å

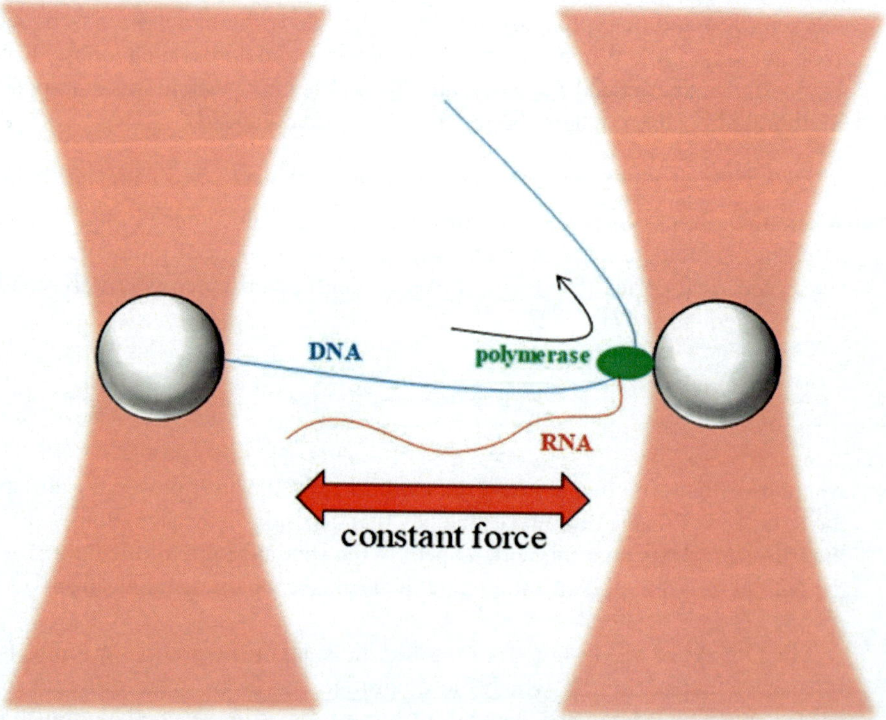

Fig. 1.4 Schematic of RNA polymerase translocation using optical trapping

were obtained, which corresponds to the length of one base in DNA. Thus, a step-by-step translocation of RNA polymerase was observed. Kellermayer et al. demonstrated folding–unfolding transitions in single titin molecules using optical tweezers [79]. Two ends of the titin molecule were anchored on two different beads, where one was entrapped by optical tweezers and the other was fixed on the tip of the micropipette. The titin molecules were stretched by the movement of the micropipette away from the optically trapped bead.

Optoelectronic tweezers (OETs) have been recently developed as an optical manipulation tool [80–82]. The illumination of light on a photoconductor substrate generates a spatially varied electric field. This ununiform electric field induces DEP, AC electroosmosis (ACEO), and electrothermal flow. An EDL is tangentially formed in the light illumination region near the electrode in an electrolyte solution. Because the counterion migration along this potential gradient is induced, an electroosmotic flow-driven vortex is caused. The time-averaged ACEO flow velocity is given by [81]:

$$< u_{\mathrm{x}} >= \frac{1}{2}\mathrm{Re}\left[\frac{\sigma_{\mathrm{q}} E_{\mathrm{t}}^{*}}{\kappa\eta}\right] \tag{1.9}$$

where σ_{q} is the charge density in the EDL, E_{t}^{*} is the complex conjugate of the electric field induced by the light, and η is the viscosity of the medium. Because an infrared laser illumination onto an indium tin oxide surface causes heating, electrothermal flow occurs. The time-averaged electrothermal force ($<F_{\mathrm{et}}>$) is represented by [81]:

$$< F_{\mathrm{et}} >= \frac{1}{2}\mathrm{Re}\left[\left(\frac{(\sigma\nabla\varepsilon - \varepsilon\nabla\sigma)\bullet E}{\sigma + i\omega\varepsilon}\right)E^{*} - \frac{1}{2}|E|^{2}\nabla\varepsilon\right] \tag{1.10}$$

where ω is the angular frequency of the AC electric field, σ and ε are the conductivity and permittivity of the medium, respectively. These three effects (DEP, ACEO, and electrothermal flow) are useful for trapping particles and macromolecules on the substrate surface.

Neale et al. used OETs to measure the relative stiffness of murine erythrocytes [83]. In this system, nonspherical cells were aligned and stretched in the direction of the electric field. A 10% change in the diameter of healthy mouse erythrocytes, which were used to evaluate relative stiffness, was detected. Chiou et al. reported the separation of live human B cells from dead cells [84]. When the cells were dispersed in a buffer with low conductivity, the difference in the conductivity between live and dead cells became evident because dead cells could not maintain an ion concentration difference between intercellular and extracellular environments. Thus, live cells moved to the bright areas, whereas dead cells were repelled from the radiation area.

Plasmonic optical tweezers

Nano-sized materials or molecules cannot be directly trapped using optical tweezers. Thus, a novel optical manipulation technique has recently been developed to overcome this limitation, [85] which is called plasmonic optical tweezers (POT). When light is illuminated onto a metal nanoparticle or thin layer, surface plasmon resonance (SPR) occurs. The optical force induced by the electromagnetic field is expressed by [86, 87]:

$$< F_{\mathrm{oe}} >= \int_{\partial V} < T(r, t) > n da \tag{1.11}$$

where ∂V is the surface of a volume enclosing an irradiated structure and T is the Maxwell stress tensor. For a particle with size a smaller than the wavelength and length of the electromagnetic field, Raleigh scattering occurs. In this case, the particle is regarded as a point dipole. The optical gradient force becomes dominant and is given by:

$$F_{\mathrm{POT}} = \frac{2\pi\alpha}{c{n_1}^2}\nabla I_0 \tag{1.12}$$

where I_0 is the intensity distribution of the electromagnetic field and α is the polarizability of the particle, which is given by:

$$\alpha = {n_1}^2 a^3 \frac{(n_2/n_1)^2 - 1}{(n_2/n_1)^2 + 2} \tag{1.13}$$

where n_2 is the refractive index of the particle. F_{POT} can be exerted on a particle with a diameter ranging from 10 nm to 1 μm [88].

Kim et al. designed nanoplasmonic structures for single DNA molecule trapping based on the POT [89]. SPR was generated by a 1050 nm laser around gold nanoholes of 400 nm diameter and 100 nm thickness. Both 4.7 kbp plasmid DNAs and 48 kbp DNAs were trapped by the POT force. Yoon et al. developed a POT device using a bowtie nanostructure, which caused SPR between the two sharp tips, for the trapping of a single sub-5-nm particle [90]. A quantum dot, trapped between the tips, enhanced the electromagnetic field and the second harmonic signal. Zhao et al. calculated the trapping behavior of chiral nanoparticles with POT [91]. The circularly polarized light was illuminated to a coaxial plasmonic aperture composed of a subwavelength silica channel embedded in a silver substrate. This allowed the stable trapping of sub-20 nm dielectric nanoparticles. Their calculations showed that enantiomers experience trapping forces of different signs and magnitudes.

Thermophoresis

Under a temperature gradient (∇T), a movement of suspended particles is induced, known as thermophoresis [92–94]. This gradient generates a flux (j), which is proportional to ∇T:

$$j = -cD_{\mathrm{T}}\nabla T \tag{1.14}$$

where c is the molecular concentration and D_{T} is the thermal diffusion coefficient. In the steady state, this thermophoretic flux is counterbalanced by mass diffusion:

$$j = -D\nabla c \tag{1.15}$$

where D is the diffusion coefficient. The Soret coefficient (S_{T}), which is widely used to quantify the strength of the thermophoretic forces, is defined as the ratio of D to D_{T}:

$$S_{\mathrm{T}} = \frac{D}{D_{\mathrm{T}}} \tag{1.16}$$

If $S_{\mathrm{T}} > 0$, the particles move toward the cold side and vice versa. The S_{T} value is affected by the factors described in the following equation:

$$S_{\mathrm{T}} = \frac{A}{kT}\left(-\Delta S_{\mathrm{hyd}}(T) + \frac{\beta\sigma_{\mathrm{eff}}^{2}}{4\varepsilon\varepsilon_{0T}}\lambda_{\mathrm{DH}}\right) \tag{1.17}$$

where A is the surface area of the molecule, σ_{eff} is the effective charge, ΔS_{hyb} is the hydration entropy, λ_{D} is the Debye length, and β is the temperature derivative of ε.

The interfacial solvent structure affects the ε of the particle and solvent itself. The local entropy change at a distance z from the particle surface is given by [95]:

$$h(z) = \frac{1}{2}\left(\varepsilon + T\frac{\partial\varepsilon}{\partial T}\right)E^{2}(z) \tag{1.18}$$

where E is the electrostatic potential in the EDL described by $E(z) = \zeta\exp(-\kappa z)$. The permittivity of water is largely affected by the particle surface and varies near the surface [96]. As a result, entropy-driven thermophoresis occurs.

Vigolo et al. reported that the migration behavior of polystyrene (PS) nanoparticles in a thermal gradient depends on the type of electrolyte. The particles in the NaCl solution moved to the hot side, whereas the particles moved to the opposite side in the NaOH solution because the S_{T} for OH^{-} is larger than that for Cl^{-}. Zhou et al. investigated the particle size effect on thermophoresis [97]. The linear relationship between the particle size and D_{T} was confirmed, which was explained by an analytical model. Batten et al. evaluated the effect of various factors on S_T [98]. They confirmed that S_T depends on T, molecular mass, and styrene ratio in the copolymer.

Thermophoresis was also used to trap particles. The mechanism of thermophoretic trapping involves the asymmetry of ε at the interface between the particle and the solvent. When the solution or the substrate is heated by laser illumination, a temperature gradient is generated around the focal point of the laser. If a particle with $S_T < 0$ is placed on the substrate, the particle is entrapped at the hot spot [95].

Hill et al. reported the manipulation of lipid vesicles using thermophoretic trapping [99]. The vesicles with 2, 3, and 4 μm in diameter were separated using multiple lasers. Lin et al. demonstrated the trapping and manipulation of living cells with laser-induced thermophoresis [100]. The cells were aligned at an arbitrary spatial interval with a resolution of 100 nm.

1.3 Particle Separation and Manipulation in an Acoustic Field

Ultrasound is also used for the separation and manipulation of particles because of its advantages as a non-contact and noninvasive force. The detailed theoretical studies of the acoustic radiation force are discussed in Chapter 2. In brief, particles in an acoustic standing wave field experience the acoustic radiation force (F_{ac}), which is given by [33, 37, 101–103]:

$$F_{ac} = -\frac{4}{3}\pi r^3 k E_{ac} A \sin(2kz) \tag{1.19}$$

$$A = \frac{5\rho^* - 2\rho}{2\rho^* + \rho} - \frac{\rho c^2}{\rho^* c^{*2}} = \frac{5\rho^* - 2\rho}{2\rho^* + \rho} - \frac{\gamma^*}{\gamma} \tag{1.20}$$

$$E_{ac} = \alpha V^2 \tag{1.21}$$

where k is the wave number of the ultrasound, z is the distance from the node or antinode of the standing wave, a is the device-dependent parameter, V is the voltage applied to the transducer, and ρ and γ are the density and compressibility of the medium, respectively (asterisk represents the parameters of the particle). The sign of F_{ac} is determined by A. $A < 0$ means the movement of a particle to the antinode of the standing wave, whereas positive A causes the movement of the particle toward the node.

Acoustic separation is classified into two techniques depending on sound propagation: bulk acoustic wave (BAW) and surface acoustic wave (SAW) [37, 103]. In typical BAW systems, a standing wave is generated by a piezoelectric transducer. In contrast, in SAW systems, one or two interdigital transducers (IDTs) on the piezoelectric substrate generate surface sound waves. The SAW system can be operated by a single traveling acoustic field (TSAW) or by a standing acoustic wave field (SSAW).

Chen et al. demonstrated the high-throughput separation of platelets from blood using a BAW system [104]. The channel with a depth of one-fourth of the standing wave wavelength was designed. Red and white blood cells were repelled to the node of the standing wave, whereas the platelets were not affected by F_{ac}. The recovery of platelets was 86.2%. Collins et al. reported a rapid single-particle sorting system based on the TSAW [105]. Highly focused TSAW with a frequency of 386 MHz was generated by focused IDTs. This system separated 2 μm particles from 1 μm particles. Li et al. have recently proposed the separation of acoustofluidic bacteria using 15°-tilted IDTs, which generates the SSAW [106]. *E. coli* bacteria flowed out of one outlet by the flow, whereas the blood cells moved along the tilted node of SSAW toward another outlet, leading to more than 96% recovery of *E. coli*.

The acoustic tweezer is one of the manipulation techniques used in various biological studies [88, 107]. The three main types of acoustic tweezers include standing wave (SWTs), traveling wave (TWTs), and acoustic streaming tweezers (ASTs). SWTs and TWTs use acoustic radiation forces exterted on particles or fluids. In contrast, particles are manipulated by acoustically induced fluid flows in AST systems.

Two different modes, BAW and SAW, are utilized for the SWT. Shi et al. demonstrated cell and microparticle patterning in two-dimensional SSAW fields using two orthogonally arranged IDTs [108]. The cells and particles were aggregated at the intersection of the node generated by each IDT. Ding et al. designed a real-time control manipulation of microparticles, cells, and organisms using four orthogonally arranged chirped IDTs [109]. The particles were manipulated by adjusting the frequency of one or two IDTs. The location of the ultrasound node generated by the two IDTs was thereby controlled. These manipulation techniques using SSAWs provide advantages such as precise particle control and simple design. However, the throughput is low, and only limited designs are feasible for manipulation using SAW techniques. Foresti et al. reported the acoustic levitation of droplets and particles in air using BAWs generated by a multi-phase transducer and found the spinning of the droplets [110]. Courtney et al. demonstrated microparticle manipulation by modulating the phase of BAWs using three piezoelectric transducers [111]. Particles levitated by the transducer placed at the bottom of the cell moved along the nodal plane by varying the phase of BAWs generated by a pair of faced transducers. The SWTs based on BAWs have high throughput, but they are not suitable for precise particle manipulation. The generation of excess heat due to high power is also a critical issue.

TWTs are also classified into active and passive methods. Active methods utilize complex acoustic beams using a single-transducer element or an array of elements. Marzo et al. demonstrated that particles were levitated and manipulated using an ultrasound transducer array by adjusting the phases of individual elements [112]. Hwang et al. used a high-frequency $LiNbO_3$ transducer with a center frequency of 193 MHz for the study of intracellular calcium signaling in SKBR-3 human breast cancer cells [113]. When a fibronectin (FNT) contacted the SKBR cell, the concentration of Ca^{2+} increased. Therefore, intracellular Ca^{2+} changes were controllable by the contact of the cell with an FNT-coated microparticle, which was acoustically entrapped. Active methods are highly flexible because the acoustic field can be simply modified, for example, by the frequency change. However, they require multiple transducers or a multiplexed transmission system.

For passive methods, acoustic metamaterials and phononic crystals are used. Bourquin et al. designed a phononic crystal structure to control the position of droplets and enhance the acoustic energy on a non-piezoelectric substrate and observed the jetting of droplets from the substrate [114]. Memoli et al. demonstrated an acoustic levitation of PS particles using a collection of metamaterial bricks on the transducer array [115]. An acoustic beam was generated in this system. Passive methods allow simple cell designs that can be easily fabricated. In contrast, because one structure generates only a few acoustic field patterns, careful simulations and calculations are needed before the fabrication of the structure.

The absorption of acoustic energy by a liquid induces a steady fluid flow, which is used for the indirect manipulation of particles. This flow, called acoustic streaming, is typically generated by oscillating microbubbles or solid structures. Yazdi et al. reported bacterial aggregation and biofilm formation based on a microbubble vibration [116]. *E. coli* bacteria were entrapped in a microchip by a vertical flow generated by microbubbles. The bubble-based AST allowed selective frequency actuation because the microbubble vibration is frequency-dependent. However, as the bubble size was not controlled in this system, the reproducibility was poor. Li et al. demonstrated silver nanowire trapping using an acoustic streaming generated by fiberglass vibration, which was induced by a sandwiched piezoelectric transducer [117]. The methods based on solid structure have high stability and reproducibility and are useful for manipulation in highly viscous liquids such as blood and sputum. However, the flexibility is low because the vibration patterns are limited.

1.4 Purpose of This Study

As described above, a number of researchers have used various external fields for particle separation and manipulation. In particular, the combination of two forces provides high performance that is not accomplished by individual forces. Masudo et al. proposed a combined acoustic-gravitational (CAG) field to separate particles [118–120]. The levitation coordinates of particles depend on their density and compressibility but are independent of particle sizes. Because microparticles have a small reaction space, only a small number of reactions induce drastic changes in their chemical and physical properties. If these changes can be measured as a levitation coordinate shift, the extent of the reaction occurring on the surface of particles can be quantitatively evaluated. The dynamic evaluation of the reaction is also possible based on the time dependence of the levitation behavior in the CAG field.

Recently, Whitesides et al. reported that the reaction rate of free radicals in a microdroplet can be determined from its levitation coordinate change using MagLev [59]. However, MagLev utilization is limited because it is applicable to mm-sized particles but inapplicable to smaller dimensions. Suwa and Watarai et al. evaluated the adsorption of a complex of dysprosium ions with lauric acid onto the surface of fluorotoluene microdroplets from magnetophoretic velocity [121–123]. Because magnetic particles or media should be used for magnetic separation, the reactions that can be evaluated are limited. Thus, only a few studies of trace analyses based on particle behavior in the external field have been reported to the best of the author's knowledge.

The purpose of this thesis is to propose a new detection scheme based on measurements of the particle levitation coordinate shift in the CAG field. Various detection concepts and schemes have been designed and demonstrated, including zeptomole (zmol) detection, label-free detection, simultaneous quantification of multiple targets, and application to real samples. The details of the proposed methods are discussed in the following sections.

References

1. Miller DL, Smith NB, Bailey MR, Czarnota GJ, Hynynen K, Makin IRS (2012) Overview of therapeutic ultrasound applications and safety considerations. J Ultrasound Med 31:623–634
2. Szreniawa-Sztajnert A, Zabiegała B, Namieśnik J (2013) Developments in ultrasound-assisted microextraction techniques for isolation and preconcentration of organic analytes from aqueous samples. TrAC Trends in Anal Chem 49:45–54
3. Wood AKW, Sehgal CM (2015) A review of low-intensity ultrasound for cancer therapy. Ultrasound Med Biol 41:905–928
4. Chemat F, Rombaut N, Sicaire A-G, Meullemiestre A, Fabiano-Tixier A-S, Abert-Vian M (2017) Ultrasound assisted extraction of food and natural products. Mechanisms, techniques, combinations, protocols and applications. A review. Ultrason Sonochem 34:540–560
5. Gessner R, Dayton PA (2010) Advances in molecular imaging with ultrasound. Mol Imaging 9:117–127
6. Klibanov AL (1999) Targeted delivery of gas-filled microspheres, contrast agents for ultrasound imaging. Adv Drug Deliver Rev 37:139–157
7. Beard P (2011) Biomedical photoacoustic imaging. Interface Focus 1:602–631
8. Dong B, Sun C, Zhang HF (2017) Optical detection of ultrasound in photoacoustic imaging. IEEE Trans Biomed Eng 64:4–15
9. Bhanvase BA, Sonawane SH (2014) Ultrasound assisted in situ emulsion polymerization for polymer nanocomposite: a review. Chem Eng Process 85:86–107
10. Kale SN, Deore SL (2016) Emulsion micro emulsion and nano emulsion: a review. Syst Rev Pharm 8:39–47
11. Chai X, Kobayashi T, Fujii N (1999) Ultrasound-associated cleaning of polymeric membranes for water treatment. Sep Purif Technol 15:139–146
12. Sauter C, Emin MA, Schuchmann HP, Tavman S (2008) Influence of hydrostatic pressure and sound amplitude on the ultrasound induced dispersion and de-agglomeration of nanoparticles. Ultrason Sonochem 15:517–523
13. Uchida T, Kikuchi T, Aoki T, Kawashima N, Takeuchi S (2009) Effects of ultrasound exposure time on nanometer-sized diamond particles dispersion. Jpn J Appl Phys 48:07GH03

14. Virot M, Szenknect S, Chave T, Dacheux N, Moisy P, Nikitenko SI (2013) Uranium carbide dissolution in nitric solution: Sonication vs. silent conditions. J Nucl Mater 441:421–430
15. Adewuyi YG (2001) Sonochemistry: environmental science and engineering applications. Ind Eng Chem Res 40:4681–4715
16. Thompson LH, Doraiswamy LK (1999) Sonochemistry: science and engineering. Ind Eng Chem Res 38:1215–1249
17. Pokhrel N, Vabbina PK, Pala N (2016) Sonochemistry: science and engineering. Ultrason Sonochem 29:104–128
18. Suslick KS, Hammerton DA, Cline JRE (1986) The sonochemical hot spot. J Am Chem Soc 108:5641–5642
19. Gole VL, Gogate PR (2012) A review on intensification of synthesis of biodiesel from sustainable feed stock using sonochemical reactors. Chem Eng Process 53:1–9
20. Xu H, Zeiger BW, Suslick KS (2013) Sonochemical synthesis of nanomaterials. Chem Soc Rev 42:2555–2567
21. Geng J, Jiang L, Zhu J (2012) Crystal formation and growth mechanism of inorganic nanomaterials in sonochemical syntheses. Sci China: Chem 55:2292–2310
22. Chowdhury P, Viraraghavan T (2009) Sonochemical degradation of chlorinated organic compounds, phenolic compounds and organic dyes—a review. Sci Total Environ 407:2474–2492
23. Sivasankar T, Moholkar VS (2009) Physical insights into the sonochemical degradation of recalcitrant organic pollutants with cavitation bubble dynamics. Ultrason Sonochem 16:769–781
24. Kost J, Leong K, Langer R (1989) Ultrasound-enhanced polymer degradation and release of incorporated substances. Proc Natl Acad Sci 86:7663–7666
25. Bruus H (2012) Acoustofluidics 7: the acoustic radiation force on small particles. Lab Chip 12:1014–1021
26. Sarvazyan AP, Rudenko OV, Nyborg WL (2010) Biomedical applications of radiation force of ultrasound: historical roots and physical basis. Ultrasound Med Biol 36:1379–1394
27. Wu J, Nyborg WL (2008) Ultrasound, cavitation bubbles and their interaction with cells. Adv Drug Deliv Rev 60:1103–1116
28. Priego-Capote F, de Castro L (2006) Ultrasound-assisted levitation: lab-on-a-drop. TrAC Trends in Anal Chem 25:856–867
29. Xie WJ, Cao CD, Lu YJ, Wei B (2002) Levitation of iridium and liquid mercury by ultrasound. Phys Rev Lett 89:104304
30. Johansson L, Enlund J, Johansson S, Katardjiev I, Yantchev V (2012) Surface acoustic wave induced particle manipulation in a PDMS channel–principle concepts for continuous flow applications. Biomed Microdevices 14:279–289
31. Bernassau AL, Ong CK, Ma Y, MacPherson PGA, Courtney CRP, Riehle M, Drinkwater BW, Cumming DRS (2011) Two-dimensional manipulation of micro particles by acoustic radiation pressure in a heptagon cell. IEEE Trans Sonics Ultrason 58:2132–2138
32. Sajeesh P, Sen AK (2013) Particle separation and sorting in microfluidic devices: a review. Microfluid Nanofluid 17:1–52
33. Lenshof A, Laurell T (2010) Continuous separation of cells and particles in microfluidic systems. Chem Soc Rev 39:1203–1217
34. Lee G-H, Kim S-H, Ahn K, Lee S-H, Park JY (2016) Separation and sorting of cells in microsystems using physical principles. J Micromech Microeng 26:013003
35. Gascoyne PRC, Vykoukal J (2002) Particle separation by dielectrophoresis. Electrophoresis 23:1973–1983
36. Kim D, Sonker M, Ros A (2019) Dielectrophoresis: from molecular to micrometer-scale analytes. Anal Chem 91:277–295
37. Tenje M, Fornell A, Ohlin M, Nilsson J (2018) Particle manipulation methods in droplet microfluidics. Anal Chem 90:1434–1443
38. Yunus NA, Nili H, Green NG (2013) Continuous separation of colloidal particles using dielectrophoresis. Electrophoresis 34:969–978

39. Jones PV, Salmon GL, Ros A (2017) Continuous separation of DNA molecules by size using insulator-based dielectrophoresis. Anal Chem 89:1531–1539
40. Han KH, Frazier AB (2008) Lateral-driven continuous dielectrophoretic microseparators for blood cells suspended in a highly conductive medium. Lab Chip 8:1079–1086
41. Zhao K, Larasati, Duncker BP, Li D (2019) Continuous cell characterization and separation by microfluidic alternating current dielectrophoresis. Anal Chem 91:6304–6314
42. Yavuz CT, Prakash A, Mayo JT, Colvin VL (2009) Magnetic separations: from steel plants to biotechnology. Chem Eng Sci 64:2510–2521
43. Borlido L, Azevedo AM, Roque AC, Aires-Barros MR (2013) Magnetic separations in biotechnology. Biotechnol Adv 31:1374–1385
44. Iranmanesh M, Hulliger J (2017) Magnetic separation: its application in mining, waste purification, medicine, biochemistry and chemistry. Chem Soc Rev 46:5925–5934
45. Brouzes E, Kruse T, Kimmerling R, Strey HH (2015) Rapid and continuous magnetic separation in droplet microfluidic devices. Lab Chip 15:908–919
46. Zhu T, Marrero F, Mao L (2010) Continuous separation of non-magnetic particles inside ferrofluids. Microfluid Nanofluid 9:1003–1009
47. Pamme N, Manz A (2004) On-chip free-flow magnetophoresis: continuous flow separation of magnetic particles and agglomerates. Anal Chem 76:7250–7256
48. Adams JD, Kim U, Soh HT (2008) Multitarget magnetic activated cell sorter. Proc Natl Acad Sci 105:18165–18170
49. Furlani EP (2007) Magnetophoretic separation of blood cells at the microscale. J Phys D Appl Phys 40:1313–1319
50. De Vlaminck I, Dekker C (2012) Recent advances in magnetic tweezers. Annu Rev Biophys 41:453–472
51. Neuman KC, Nagy A (2008) Single-molecule force spectroscopy: optical tweezers, magnetic tweezers and atomic force microscopy. Nat Methods 5:491–505
52. Mosconi F, Allemand JF, Croquette V (2011) Soft magnetic tweezers: a proof of principle. Rev Sci Instrum 82:034302
53. Gore J, Bryant Z, Stone MD, Nollman M, Cozzarelli NR, Bustamante C (2006) Mechanochemical analysis of DNA gyrase using rotor bead tracking. Nature 439:100–104
54. Itoh H, Takahashi A, Adachi K, Noji H, Yasuda R, Yoshida M, Kinosita K Jr (2004) Mechanically driven ATP synthesis by F1-ATPase. Nature 427:465–468
55. Lipfert J, Skinner GM, Keegstra JM, Hensgens T, Jager E, Dulin D, Kober M, Yu Z, Donkers SP, Chou FC, Das R, Dekker NH (2014) Double-stranded RNA under force and torque: similarities to and striking differences from double-stranded DNA. Proc Natl Acad Sci 111:15408–15413
56. Harada Y, Ohara O, Takatsuki A, Itoh H, Shimamoto N, Kinosita K Jr (2001) Direct observation of DNA rotation during transcription by escherichia coli RNA polymerase. Nature 409:113–115
57. Ge S, Whitesides GM (2018) Axial" magnetic levitation using ring magnets enables simple density-based analysis, separation, and manipulation. Anal Chem 90:12239–12245
58. Nemiroski A, Kumar AA, Soh S, Harburg DV, Yu HD, Whitesides GM (2016) High-sensitivity measurement of density by magnetic levitation. Anal Chem 88:2666–2674
59. Ge S, Semenov SN, Nagarkar AA, Milette J, Christodouleas DC, Yuan L, Whitesides GM (2017) Magnetic levitation to characterize the kinetics of free-radical polymerization. J Am Chem Soc 139:18688–18697
60. Mirica KA, Shevkoplyas SS, Phillips ST, Gupta M, Whitesides GM (2009) Measuring densities of solids and liquids using magnetic levitation: fundamentals. J Am Chem Soc 131:10049–10058
61. Bwambok DK, Thuo MM, Atkinson MB, Mirica KA, Shapiro ND, Whitesides GM (2013) Paramagnetic ionic liquids for measurements of density using magnetic levitation. Anal Chem 85:8442–8447
62. Kremser L, Blaas D, Kenndler E (2004) Capillary electrophoresis of biological particles: viruses, bacteria, and eukaryotic cells. Electrophoresis 25:2282–2291

63. Radko SP, Chrambach A (2002) Separation and characterization of sub-μm- and μm-sized particles by capillary zone electrophoresis. Electrophoresis 23:1957–1972
64. Surugau N, Urban PL (2009) Electrophoretic methods for separation of nanoparticles. J Sep Sci 32:1889–1906
65. Okun V, Ronacher B, Blaas D, Kenndler E (2000) Capillary electrophoresis with postcolumn infectivity assay for the analysis of different serotypes of human rhinovirus. Anal Chem 72:2553–2558
66. Sonohara R, Muromatsu N, Ohshima H, Kondo T (1995) Difference in surface properties between escherichia coli and staphylococcus aureus as revealed by electrophoretic mobility measurements. Biophys Chem 55:273–277
67. Tsukagoshi K, Yamaguchi A, Morihara N, Nakajima R, Murata M, Maeda M (2001) Migration behavior of the polymer particles possessing carboxylate and borate moieties in capillary electrophoresis. Chem Lett 30:926–927
68. Armstrong DW, Schulte G, Schneiderheinze JM, Westenberg JD (1999) Separating microbes in the manner of molecules. 1. Capillary electrokinetic approaches. Anal Chem 71:5465–5469
69. Cohen AD, Moerner WE (2005) Method for trapping and manipulating nanoscale objects in solution. Appl Phys Lett 86:093109
70. Yazbeck R, Alibakhshi MA, Von Schoppe J, Ekinci KL, Duan C (2019) Characterization and manipulation of single nanoparticles using a nanopore-based electrokinetic tweezer. Nanoscale 11:22924–22931
71. Jonas A, Zemanek P (2008) Light at work: the use of optical forces for particle manipulation, sorting, and analysis. Electrophoresis 29:4813–4851
72. Dholakia K, MacDonald MP, Zemánek P, Čižmár T (2007) Cellular and colloidal separation using optical forces. Methods Cell Biol 82:467–495
73. Levison HF, Morbidelli A (2003) The formation of the Kuiper belt by the outward transport of bodies during Neptune's migration. Nature 426:419–421
74. Imasaka T, Kawabata Y, Kaneta T, Ishidzu Y (1995) Optical chromatography. Anal Chem 67:1763–1765
75. Hebert CG, Hart S, Leski TA, Terray A, Lu Q (2017) Label-free detection of bacillus anthracis spore uptake in macrophage cells using analytical optical force measurements. Anal Chem 89:10296–10302
76. Ashkin A (1970) Acceleration and trapping of particles by radiation pressure. Phys Rev Lett 24:156–159
77. Mammen M, Helmerson K, Kishore R, Choi SK, Phillips WD, Whitesides GM (1996) Optically controlled collisions of biological objects to evaluate potent polyvalent inhibitors of virus-cell adhesion. Chem Biol 3:757–763
78. Abbondanzieri EA, Greenleaf WJ, Shaevitz JW, Landick R, Block SM (2005) Direct observation of base-pair stepping by RNA polymerase. Nature 438:460–465
79. Kellermayer MSZ, Smith SB, Granzier HL, Bustamante C (1997) Folding-unfolding transitions in single titin molecules characterized with laser tweezers. Science 276:1112–1116
80. Wu MC (2011) Optoelectronic tweezers. Nat Photonics 5:322–324
81. Mishra A, Kwon JS, Thakur R, Wereley S (2014) Optoelectrical microfluidics as a promising tool in biology. Trends Biotechnol 32:414–421
82. Hwang H, Park JK (2011) Optoelectrofluidic platforms for chemistry and biology. Lab Chip 11:33–47
83. Neale SL, Spalding GC, Mody N, Selman C, Cooper JM (2012) Optoelectronic tweezers for the measurement of the relative stiffness of erythrocytes. Proc SPIE 8458:845827
84. Chiou PY, Ohta AT, Wu MC (2005) Massively parallel manipulation of single cells and microparticles using optical images. Nature 436:370–372
85. Grigorenko AN, Roberts NW, Dickinson MR, Zhang Y (2008) Nanometric optical tweezers based on nanostructured substrates. Nat Photonics 2:365–370
86. Huang JS, Yang YT (2015) Origin and future of plasmonic optical tweezers. Nanomaterials (Basel) 5:1048–1065

87. Han X, Sun C (2019) Plasmonic tweezers towards biomolecular and biomedical applications. Appl Sci 9:3596
88. Ozcelik A, Rufo J, Guo F, Gu Y, Li P, Lata J, Huang TJ (2018) Acoustic tweezers for the life sciences. Nat Methods 15:1021–1028
89. Kim JD, Lee YG (2014) Trapping of a single DNA molecule using nanoplasmonic structures for biosensor applications. Biomed Opt Express 5:2471–2480
90. Yoon SJ, Lee J, Han S, Kim CK, Ahn CW, Kim MK, Lee YH (2018) Non-fluorescent nanoscopic monitoring of a single trapped nanoparticle via nonlinear point sources. Nat Commun 9:2218
91. Zhao Y, Saleh AAE, Dionne JA (2016) Enantioselective optical trapping of chiral nanoparticles with plasmonic tweezers. ACS Photonics 3:304–309
92. Seidel SA, Dijkman PM, Lea WA, van den Bogaart G, Jerabek-Willemsen M, Lazic A, Joseph JS, Srinivasan P, Baaske P, Simeonov A, Katritch I, Melo FA, Ladbury JE, Schreiber G, Watts A, Braun D, Duhr S (2013) Microscale thermophoresis quantifies biomolecular interactions under previously challenging conditions. Methods 59:301–315
93. Asmari M, Ratih R, Alhazmi HA, El Deeb S (2018) Thermophoresis for characterizing biomolecular interaction. Methods 146:107–119
94. Burelbach J, Frenkel D, Pagonabarraga I, Eiser E (2018) A unified description of colloidal thermophoresis. Eur Phys J E Soft Matter 41:7
95. Lin L, Peng X, Mao Z, Wei X, Xie C, Zheng Y (2017) Interfacial-entropy-driven thermophoretic tweezers. Lab Chip 17:3061–3070
96. Peng X, Lin L, Hill EH, Kunal P, Humphrey SM, Zheng Y (2018) Optothermophoretic manipulation of colloidal particles in nonionic liquids. J Phys Chem C 122:24226–24234
97. Zhou Y, Yang C, Lam YC, Huang X (2016) Thermophoresis of charged colloidal particles in aqueous media—effect of particle size. Int J Heat Mass Transfer 101:1283–1291
98. Batten CV, Hoyos M, Martin M (1997) Thermal field-flow fractionation of colloidal materials: methylmethacrylate-styrene linear di-block copolymers. Chromatographia 45:121–126
99. Hill EH, Li J, Lin L, Liu Y, Zheng Y (2018) Opto-thermophoretic attraction, trapping, and dynamic manipulation of lipid vesicles. Langmuir 34:13252–13262
100. Lin L, Peng X, Wei X, Mao Z, Xie C, Zheng Y (2017) Thermophoretic tweezers for low-power and versatile manipulation of biological cells. ACS Nano 11:3147–3154
101. Lenshof A, Magnusson C, Laurell T (2012) Acoustofluidics 8: applications of acoustophoresis in continuous flow microsystems. Lab Chip 12:1210–1223
102. Laurell T, Petersson F, Nilsson A (2007) Chip integrated strategies for acoustic separation and manipulation of cells and particles. Chem Soc Rev 36:492–506
103. Wu M, Ozcelik A, Rufo J, Wang Z, Fang R, Jun Huang T (2019) Acoustofluidic separation of cells and particles. Microsyst Nanoeng 5:32
104. Chen S, Liu J, Liu Y, Su H, Hong Y, Jim CKW, Kwok RTK, Zhao N, Qin W, Lam JWY, Wong KS, Tang BZ (2012) An AIE-active hemicyanine fluorogen with stimuli-responsive red/blue emission: extending the pH sensing range by "switch + knob" effect. Chem Sci 3:1804–1809
105. Collins DJ, Neild A, Ai Y (2016) Highly focused high-frequency travelling surface acoustic waves (SAW) for rapid single-particle sorting. Lab Chip 16:471–479
106. Li S, Ma F, Bachman H, Cameron CE, Zeng X, Huang TJ (2017) Acoustofluidic bacteria separation. J Micromech Microeng 27:015031
107. Meng L, Cai F, Li F, Zhou W, Niu L, Zheng H (2019) Acoustic tweezers. J Phys D: Appl Phys 52:273001
108. Shi J, Ahmed D, Mao X, Lin SC, Lawit A, Huang TJ (2009) Acoustic tweezers: patterning cells and microparticles using standing surface acoustic waves (SSAW). Lab Chip 9:2890–2895
109. Ding X, Lin SC, Kiraly B, Yue H, Li S, Chiang IK, Shi J, Benkovic SJ, Huang TJ (2012) On-chip manipulation of single microparticles, cells, and organisms using surface acoustic waves. Proc Natl Acad Sci 109:11105–11109
110. Foresti D, Poulikakos D (2014) Acoustophoretic contactless elevation, orbital transport and spinning of matter in air. Phys Rev Lett 112:024301

111. Courtney CR, Ong CK, Drinkwater BW, Wilcox PD, Demore C, Cochran S, Glynne-Jones P, Hill M (2010) Manipulation of microparticles using phase-controllable ultrasonic standing waves. J Acoust Soc Am 128:195–199
112. Marzo A, Seah SA, Drinkwater BW, Sahoo DR, Long B, Subramanian S (2015) Holographic acoustic elements for manipulation of levitated objects. Nat Commun 6:8661
113. Hwang JY, Yoon CW, Lim HG, Park JM, Yoon S, Lee J, Shung KK (2015) Acoustic tweezers for studying intracellular calcium signaling in SKBR-3 human breast cancer cells. Ultrasonics 63:94–101
114. Bourquin Y, Wilson R, Zhang Y, Reboud J, Cooper JM (2011) Phononic crystals for shaping fluids. Adv Mater 23:1458–1462
115. Memoli G, Caleap M, Asakawa M, Sahoo DR, Drinkwater BW, Subramanian S (2017) Metamaterial bricks and quantization of meta-surfaces. Nat Commun 8:14608
116. Yazdi S, Ardekani AM (2012) Bacterial aggregation and biofilm formation in a vortical flow. Biomicrofluidics 6:044114
117. Li N, Hu J, Li H, Bhuyan S, Zhou Y (2012) Mobile acoustic streaming based trapping and 3-dimensional transfer of a single nanowire. Appl Phys Lett 101:093113
118. Masudo T, Okada T (2001) Particle characterization and separation by a coupled acoustic—gravity field. Anal Chem 73:3467–3471
119. Kanazaki T, Okada T (2012) Two-dimensional particle separation in coupled acoustic-gravity-flow field vertically by composition and laterally by size. Anal Chem 84:10750–10755
120. Kanazaki T, Hirawa S, Harada M, Okada T (2010) Coupled acoustic-gravity field for dynamic evaluation of ion exchange with a single resin bead. Anal Chem 82:4472–4478
121. Funaki M, Suwa M, Watarai H (2017) Electromagnetophoretic micro-convection around a droplet in a capillary. Anal Sci 33:1013–1019
122. Suwa M, Watarai H (2008) Magnetophoretic evaluation of interfacial adsorption of dysprosium(III) on a single microdroplet. Anal Sci 24:133–137
123. Watarai H, Duc HTT, Lan TTN, Zhang T, Tsukahara S (2014) Zero-velocity magnetophoretic method for the determination of particle magnetic susceptibility. Anal Sci 30:745–749

Chapter 2
Theory of Combined Acoustic-Gravitational Field

Abstract In this chapter, the principle of an acoustic radiation force and a combined acoustic-gravitational (CAG) field is discussed. From Sect. 2.1 to Sect. 2.4, the mathematical interpretation of the acoustic radiation force is described. Based on this, the principle of the CAG field is constructed in Sect. 2.5.

Keywords Acoustic radiation force · Gravitational field · Sedimentation force · Equilibrium levitation coordinate

2.1 Radiation Force on a Moving Boundary

The pressure fluctuation in a perfect fluid, where the acoustic wave is propagating, is represented by:

$$\delta p = \rho_0\left(\dot{\Phi}_1 + \dot{\Phi}_2\right) - \frac{1}{2}\rho_0 u_1^2 + \frac{1}{2}\frac{\rho_0}{c^2}\dot{\Phi}_1^2 \tag{2.1}$$

where ρ_0 is the equilibrium density of the fluid, c is the acoustic velocity in the medium, u_1 is the first-order particle velocity ($= -\Phi_1$), Φ_1 is the first-order velocity potential given by the approximated wave equation $\Delta\Phi_1 = (1 / c^2)\ \partial^2\Phi_1 / \partial t^2$, and Φ_2 is the second-order velocity potential.

Assuming that a closed boundary **S** in the medium is moving with a small velocity u_1, the position at time t is represented as $\mathbf{S}(t)$. The acoustic radiation force on the moving boundary $< P >$ is represented by:

$$\begin{aligned}\langle p\rangle &= \left\langle -\iint_{s(t)} \delta p \mathbf{n} df \right\rangle = \left\langle -\iint_{s(t)} \rho_0 \dot{\Phi}_1 \mathbf{n} df \right\rangle + \left\langle -\iint_{s_0} \rho_0 \dot{\Phi}_2 \mathbf{n} df \right\rangle \\ &+ \left\langle -\iint_{s_0} \frac{1}{2}\rho_0 u_1^2 \mathbf{n} df \right\rangle + \left\langle -\iint_{s_0} \frac{1}{2}\frac{\rho_0}{c^2}\dot{\Phi}_1^2 \mathbf{n} df \right\rangle \end{aligned} \tag{2.2}$$

where S_0 is the boundary of the equilibrium position and **n** is an outward nominal to $\mathbf{S}(t)$ [1, 2].

© The Author(s), under exclusive license to Springer Nature Singapore Pte Ltd. 2021

A. Miyagawa, *Acoustic Levitation-Based Trace-Level Biosensing*, Springer Theses,
https://doi.org/10.1007/978-981-16-1425-5_2

When a steady state is established in the primary and secondary acoustic fields, the second-order velocity, u_2, and $\rho_0\dot{\Phi}_2$ are represented by:

$$u_2 = -\nabla\Phi_2 = \overrightarrow{const} + \overrightarrow{f_0} + \sum_{n=1}\left\{\overrightarrow{f_{1,n}} \cos n\omega t + \overrightarrow{f_{2,n}} \sin n\omega t\right\} \tag{2.3}$$

$$\rho_0\dot{\Phi}_2 = \delta p_2 + \frac{1}{2}\rho_0 u_1^2 - \frac{1}{2}\frac{\rho_0}{c^2}\dot{\Phi}_1^2 = const + g_0 + \sum_{n=1}\{g_{1,n} \cos n\omega t + g_{2,n} \sin n\omega t\} \tag{2.4}$$

where δp_2 is the second-order pressure variation and $\overrightarrow{f_0}$, $\overrightarrow{f_{1,n}}$, $\overrightarrow{f_{2,n}}$, g_0, $g_{1,n}$, $g_{2,n}$ are the functions of the coordinate.

From Eqs. (2.3) and (2.4), we obtain:

$$g_0(x, y, z) = const \tag{2.5}$$

then

$$\left\langle \iint_{S_0} \rho_0\dot{\Phi}_2 \boldsymbol{n} df = 0 \right\rangle \tag{2.6}$$

By expressing the average velocity on the boundary as $u_n\mathbf{n}$, the following equation is obtained:

$$\frac{d}{dt}\left\{\iint_{S(t)} \rho_0\Phi_1 \boldsymbol{n} df\right\} = \iint_{S(t)} \rho_0\dot{\Phi}_1 \boldsymbol{n} df + \iint_{S(t)} \rho_0\nabla\Phi_1 \cdot u_n df \tag{2.7}$$

Because both $\boldsymbol{\Phi}_1$ and the displacement of the boundary periodically change with respect to time, by taking the time average of Eq. (2.7) and rewriting $\boldsymbol{S}(t)$ to $\boldsymbol{S}_0$ in the last integration symbol, we obtain:

$$\left\langle \iint_{S(t)} \rho_0\dot{\Phi}_1 \boldsymbol{n} df \right\rangle = \left\langle -\iint_{S_0} \rho_0 \boldsymbol{\nabla\Phi}_1 \cdot u_n df \right\rangle = \left\langle \iint_{S_0} \rho_0(u_n\mathbf{n} + u_t\mathbf{t})u_n df \right\rangle \tag{2.8}$$

where $u_t\mathbf{t}$ is a tangential particle velocity in the medium.

Substituting Eqs. (2.8) and (2.6) into Eq. (2.2) yields:

$$\langle P \rangle = \left\langle -\iint_{S_0} \rho_0(u_n\boldsymbol{n} + u_t\boldsymbol{t})u_n df \right\rangle + \left\langle \iint_{S_0} \frac{1}{2}\rho_0 u_1^2 \boldsymbol{n} df \right\rangle - \left\langle \iint_{S_0} \frac{1}{2}\frac{\rho_0}{c^2}\dot{\Phi}_1^2 \boldsymbol{n} df \right\rangle \tag{2.9}$$

Thus, the radiation force on the moving boundary is calculated with respect to the first-order velocity potential. The entire surface at its equilibrium position is integrated.

If the boundary is a sphere with radius r and the acoustic field is axisymmetric with respect to $\theta = 0$, the radiation pressure along $\theta = 0$ is given by:

$$\langle P \rangle = \langle P_{uv} \rangle + \langle P_{u\theta} \rangle + \langle P_{uv,u\vartheta} \rangle + \langle P_{\Phi} \rangle \tag{2.10}$$

where

$$\langle P_{uv} \rangle = \left\langle -\pi a^2 \rho_0 \int_0^{\pi} \left(\frac{\partial \Phi_1}{\partial v} \right)^2_{v=a} \sin\theta \cos\theta d\theta \right\rangle \tag{2.11}$$

$$\langle P_{u\theta} \rangle = \left\langle -\pi \rho_0 \int_0^{\pi} \left(\frac{\partial \Phi_1}{\partial v} \right)^2_{v=a} \sin\theta \cos\theta d\theta \right\rangle \tag{2.12}$$

$$\langle P_{uv,u\theta} \rangle = \left\langle 2\pi a \lambda \rho_0 \int_0^{\pi} \left(\frac{\partial \Phi_1}{\partial v} \right)_{v=a} \left(\frac{\partial \Phi_1}{\partial \theta} \right)_{v=a} \sin^2\theta d\theta \right\rangle \tag{2.13}$$

$$\langle P_{\Phi} \rangle = \left\langle -\frac{\pi a^2 \rho_0}{c^2} \int_0^{\pi} (\dot{\Phi}_1)^2_{v=a} \sin\theta \cos\theta d\theta \right\rangle \tag{2.14}$$

Φ_1 is the solution of the scattering on a compressible sphere that satisfies the boundary condition at its equilibrium position.

2.2 Scattering of Plane Waves on a Compressible Sphere

2.2.1 *Plane Progressive Wave*

A sphere with radius a is freely placed at the origin of the Cartesian coordinate system. The incident velocity potential Φ_1 of a plane traveling wave propagating at $z = 0$ or $\theta = 0$ is given by the following equation:

$$\Phi_i = e^{i(\omega t - kz)} = e^{i\omega t} \sum_{n=0}^{\infty} (2n+1)(-i)^n J_n(kv) P_n(\cos\theta) \tag{2.15}$$

where k is the wave number, J_n is the nth Bessel series, and P_n is the nth Legendre function.

The velocity potential outside the sphere, Φ_1, is given by:

$$\Phi_1 = \Phi_i + \Phi_s \tag{2.16}$$

where Φ_i and Φ_s are the velocity potentials of the incident and scattered waves, respectively. Φ_s is expressed as:

$$\Phi_s = e^{i\omega t} \sum_{n=0}^{\infty} (2n+1)(-i)^n A_n H_n^{(2)}(kv) P_n(\cos\theta) \tag{2.17}$$

where $H_n^{(2)}$ is a second-order spherical Hankel function and A_n is a constant determined by the boundary conditions.

The velocity potential Φ^* inside the sphere is given by the following equation:

$$\Phi^* = e^{i\omega t} \sum_{n=0}^{\infty} (2n+1)(-i)^n B_n J_n(k^* v) P_n(\cos\theta) \tag{2.18}$$

where k^* is the wave number inside the sphere, the superscript * indicates the quantities concerning the sphere, and B_n is a constant to be determined together with A_n by the following boundary conditions:

$$\frac{\partial \Phi_1}{\partial v} = \frac{\partial \Phi^*}{\partial v} \quad \text{at} \quad v = a \tag{2.19}$$

$$\rho_0 \dot{\Phi}_1 = \rho^* \dot{\Phi}^* \quad \text{at} \quad v = a \tag{2.20}$$

ρ_0 and ρ^* are the densities inside and outside the medium, respectively. A_n and B_n are expressed as follows:

$$A_n = \frac{R_d k J_n(k^* a) J_n'(ka) - k^* J_n'(k^* a) J_n(ka)}{k^* J_n'(k^* a) H_n^{(2)}(ka) - R_d k J_n(k^* a) H_n^{(2)\prime}(ka)} \tag{2.21}$$

$$B_n = \frac{ik}{(ka)^2 \left[k^* J_n'(k^* a) H_n^{(2)}(ka) - R_d k J_n(k^* a) H_n^{(2)}(ka) \right]} \tag{2.22}$$

where $R_d = \rho^*/\rho_0$.

In the case of $(ka)^2 < < 1$, $(k^*a)^2 < < 1$, and $R_d = O\{(ka)\}$, B_0 is given by the following equation:

$$B_0 = \frac{3i}{-(k^* a)^2 ka + i\left[3R_d - (k^* a)^2\right]} \tag{2.23}$$

This is the case of levitated bubbles in a liquid and B_0 is proportional to the amplitude of the wave at the surface. Eq. (2.23) represents wave resonance, which occurs when:

$$\left(k^*a\right)^2 = 3R_d \tag{2.24}$$

The resonant frequency f of ideal gas bubbles is expressed by:

$$f = \frac{1}{2\pi a}\sqrt{\frac{3\gamma p_0}{\rho_0}} \tag{2.25}$$

where γ is the ratio of the specific heat of the gas and $p_0$is the hydrostatic pressure.

2.2.2 Plane Standing Wave

Assuming that the center of the sphere exists at a position away from the nodal plane by a distance h, the radiation velocity potential referred to as the equilibrium position of the center is given by:

$$\Phi_i = e^{i\omega t}\left[e^{ik(z+h)} + e^{-ik(z+h)}\right] = e^{i\omega t}\sum_{n=0}^{\infty}(2n+1)(-i)^n\delta_n J_n(kv)P_n(\cos\theta) \tag{2.26}$$

where

$$\delta_n = (-1)^n e^{ikh} + e^{-ikh} \tag{2.27}$$

The scattering velocity potential $\boldsymbol{\Phi}_s$ and the sphere internal velocity potential, $\boldsymbol{\Phi}^*$, are expressed as follows:

$$\Phi_s = e^{i\omega t}\sum_{n=0}^{\infty}(2n+1)(-i)^n C_n H_n^{(2)}(kv)P_n(\cos\theta) \tag{2.28}$$

$$\Phi^* = e^{i\omega t}\sum_{n=0}^{\infty}(2n+1)(-i)^n D_n J_n\left(k^*v\right)P_n(\cos\theta) \tag{2.29}$$

$C_n$and D_n are determined by the boundary conditions given by Eqs. (2.19) and (2.20) as:

$$C_n = A_n\boldsymbol{\delta}_n \tag{2.30}$$

$$D_n = B_n \delta_n \tag{2.31}$$

2.3 Radiation Pressure on a Compressible Sphere

It is convenient to express P_{uv}, $P_{u\theta}$, $P_{uv,\,u\theta}$, and P_{Φ} as a function of Φ^*:

$$\frac{\partial \Phi_1}{\partial v} = \frac{\partial \Phi^*}{\partial v} \quad \text{at} \quad v = a \tag{2.32}$$

$$\rho_0 \dot{\Phi}_1 = \rho^* \dot{\Phi}^* \quad \text{at} \quad v = a \tag{2.33}$$

$$\rho_0 \frac{\partial \Phi_1}{\partial \theta} = \rho^* \frac{\partial \Phi^*}{\partial \theta} \quad \text{at} \quad v = a \tag{2.34}$$

The following equations are derived from Eqs. (2.11), (2.12), (2.13), and (2.14) by substituting $\cos\theta = \mu$:

$$\langle P_{uv} \rangle = \left\langle -\pi a^2 \rho_0 \int_{-1}^{+1} \left(\frac{\partial \Phi^*}{\partial v} \right)^2_{v=a} \mu d\mu \right\rangle \tag{2.35}$$

$$\langle P_{u\theta} \rangle = \left\langle \pi R_d \rho_0 \int_{-1}^{+1} \left(\frac{\partial \Phi^*}{\partial \mu} \right)^2_{v=a} \mu (1 - \mu^2) d\mu \right\rangle \tag{2.36}$$

$$\langle P_{uv,u\theta} \rangle = \left\langle -2\pi a R_d \rho_0 \int_{-1}^{+1} \left(\frac{\partial \Phi^*}{\partial v} \right)_{v=a} \left(\frac{\partial \Phi^*}{\partial \mu} \right)_{v=a} (1 - \mu^2) d\mu \right\rangle \tag{2.37}$$

$$\langle P_{\Phi} \rangle = \left\langle -\frac{\pi a^2 R_d^2}{c^2} \rho_0 \int_{-1}^{+1} (\dot{\Phi}^*)^2_{r=a} \mu d\mu \right\rangle \tag{2.38}$$

It is convenient to write the results obtained in Sect. 2.2 in the following forms:

$$\mathrm{Re}(\Phi^*)_{v=a} = \sum_{n=0}^{\infty} (2n + 1) M_n P_n(\cos\theta) \tag{2.39}$$

$$\mathrm{Re}\left(\frac{\partial \Phi^*}{\partial v} \right)_{v=a} = \sum_{n=0}^{\infty} (2n + 1) K_n P_n(\cos\theta) \tag{2.40}$$

or

$$M_n = \mathrm{Re}\left[e^{i\omega t} (-i)^n B_n J_n(k^* a) \right] \tag{2.41}$$

$$K_n = \mathrm{Re}\left[e^{i\omega t}(-i)^n B_n k^* J_n'\left(k^* a\right)\right] \tag{2.42}$$

By substituting Eqs. (2.39) and (2.40) into Eqs. (2.35), (2.36), (2.37), and (2.38), we can express the radiation pressure in terms of M_n and K_n factors:

$$\langle P\rangle = \langle P_{uv}\rangle + \langle P_{u\theta}\rangle + \left\langle P_{uv,u\theta}\right\rangle + \langle P_{\Phi}\rangle \tag{2.43}$$

$$\langle P_{uv}\rangle = -2\pi a^2 \rho_0 \sum_{n=0}^{\infty} 2(n+1)\langle K_n K_{n+1}\rangle \tag{2.44}$$

$$\langle P_{u\theta}\rangle = 2\pi R_d^2 \rho_0 \sum_{n=0}^{\infty} 2n(n+1)(n+2)\langle M_n M_{n+1}\rangle \tag{2.45}$$

$$\begin{aligned}\left\langle P_{uv,u\theta}\right\rangle = & -2\pi a R_d \rho_0 \sum_{n=0}^{\infty} 2(n+1)(n+2)\langle K_n M_{n+1}\rangle \\ & + 2\pi a R_d \rho_0 \sum_{n=0}^{\infty} 2n(n+1)\langle M_n K_{n+1}\rangle\end{aligned} \tag{2.46}$$

$$\langle P_{\Phi}\rangle = -2\pi k^2 a^2 R_d^2 \rho_0 \sum_{n=0}^{\infty} 2(n+1)\langle M_n M_{n+1}\rangle \tag{2.47}$$

M_n and K_n are written as:

$$M_n = M_{1,n}\cos\omega t + M_{2,n}\sin\omega t \tag{2.48}$$

$$K_n = K_{1,n}\cos\omega t + K_{2,n}\sin\omega t \tag{2.49}$$

then

$$2\langle M_n M_{n+1}\rangle = M_{1,n}M_{1,n+1} + M_{2,n}M_{2,n+1} \tag{2.50}$$

$$2\langle K_n M_{n+1}\rangle = K_{1,n}M_{1,n+1} + K_{2,n}M_{2,n+1} \tag{2.51}$$

B_n in a plane propagating wave is expressed as:

$$B_n = \frac{1}{ka}\cdot\frac{j}{D_{1,n} - jD_{2,n}} \tag{2.52}$$

where

$$D_{1,n} = k^* a J_n'\left(k^* a\right) J_n(ka) - R_d ka J_n\left(k^* a\right) J_n'(ka) \tag{2.53}$$

$$D_{2,n} = k^* a J_n'(k^* a) N_n(ka) - R_d ka J_n(k^* a) N_n'(ka) \tag{2.54}$$

N_n is a spherical Neumann function. By substituting Eqs. (2.41), (2.42), and (2.52) into Eqs. (2.43), (2.44), (2.45), (2.46), and (2.47), we obtain:

$$\langle P \rangle = 2\pi \rho_0 \sum_{n=0}^{\infty} \frac{(n+1)\{D_{1,n} D_{2,n+1} - D_{2,n} D_{1,n+1}\}^2}{\{D_{1,n}^2 + D_{2,n}^2\}\{D_{1,n+1}^2 + D_{2,n+1}^2\}} \tag{2.55}$$

Therefore, the radiation pressure on the compressible sphere in a plane propagating wave field is always positive. The pressure is applied in the direction of wave propagation.

2.4 Acoustic Force on Small Particles

The radiation pressure in a microsphere satisfying $[(ka)^2, (k^*a)^2 \ll 1]$、$R_d = \rho^* / \rho_0 = O\{1\}$ is the case of a small compressible particle suspended in liquid media. If $(ka)^2$ and $(k^*a)^2$ are negligibly small compared with the sum of M_n and K_n for $n \geq 3$, they do not contribute to the preceding result. Thus, we obtain the following equation derived from Eqs. (2.43), (2.44), (2.45), and (2.46):

$$\langle P \rangle = 2\pi \rho_0 \left[2a^2(1+2R_d)\langle K_0 K_1 \rangle + 4a^2(R_d - 1)\langle K_1 K_2 \rangle + 2k^2 a^2 R_d^2 \langle M_0 M_1 \rangle + 12 R_d \langle M_1 M_2 \rangle\right] \tag{2.56}$$

2.4.1 Plane Progressive Wave

Substituting M_n and K_n into Eq. (2.56), we obtain:

$$\langle P \rangle = 2\pi \rho_0 (ka)^6 F(R_d, \sigma) \tag{2.57}$$

$$= \pi a^4 \cdot 4(ka)^4 \frac{I}{c} F(R_d, \sigma) \tag{2.58}$$

$$F(R_d, \sigma) = \frac{1}{(1+2R_d)^2}\left[\left(R_d - \frac{1+2R_d}{3R_d \sigma^2}\right)^2 + \frac{2}{9}(1-R_d)^2\right] \tag{2.59}$$

where $\sigma = c^*/c = k/k^*$, $I(=\rho_0 c k^2/2)$ is the intensity of incident waves, $F(R_d, \sigma)$ is a function of the density and compressibility and converges to the King's density factor when $\sigma \rightarrow \infty$ [3, 4]. This density–compressibility function is positive and the radiation pressure acts in the direction of wave propagation.

2.4.2 *Plane Stationary Wave*

Substituting M_n and K_n into Eq. (2.56), we obtain:

$$\langle P \rangle = 4\pi \rho_0 (ka)^3 \sin(2kh F(R_d, \sigma)) \tag{2.60}$$

$$= \pi a^2 4ka E_{\mathrm{ac}} \sin(2kh F(R_d, \sigma)) \tag{2.61}$$

where E_{ac} (=$\rho_0 k^2$) is the average ultrasound energy density in the acoustic standing wave and $F(R_d, \sigma)$ is the density–compressibility factor given by

$$F(R_d, \sigma) = \frac{R_d + [2(R_d - 1)/3]}{1 + 2R_d} - \frac{1}{3R_d \sigma^2} \tag{2.62}$$

Equation (2.62) converges to the King's density factor [3, 4] when $\sigma \rightarrow \infty$. If $F(R_d, \sigma)$ is positive, the radiation pressure urges the sphere away from the antinode, and if negative, away from the node.

2.5 Combined Acoustic-Gravitational Field

From Eqs. (2.61) and (2.62), when a particle with radius r exists in an ultrasound standing wave, the particle experiences the following acoustic radiation force (F_{ac}):

$$F_{\mathrm{ac}} = -\frac{4}{3}\pi r^3 k E_{\mathrm{ac}} A \sin(2kz) \tag{2.63}$$

where k is the wave number of the ultrasound and E_{ac} is the average ultrasound energy density. A, which is a function of the density and compressibility of a particle and medium, is represented by:

$$A = \frac{5\rho^* - 2\rho}{2\rho^* + \rho} - \frac{\rho c^2}{\rho^* c^{*2}} = \frac{5\rho^* - 2\rho}{2\rho^* + \rho} - \frac{\gamma^*}{\gamma} \tag{2.64}$$

where ρ, c, and γ are the density, velocity, and compressibility of the medium, respectively. The superscript * denotes the property of the particle.

Figure 2.1 shows the relationships between the force exerted on a particle and the

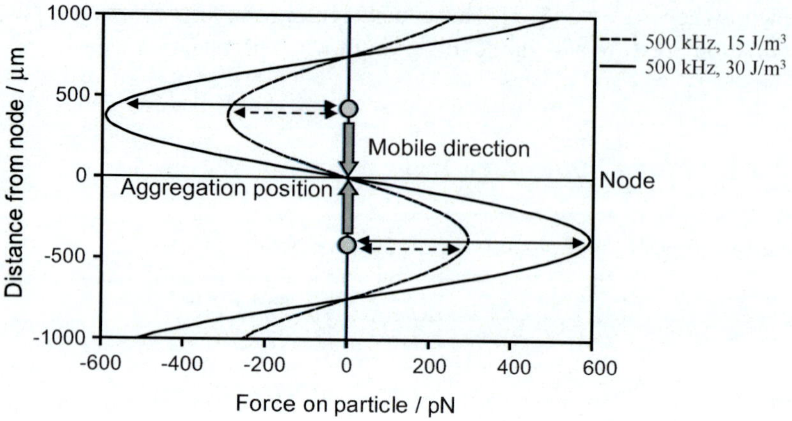

Fig. 2.1 Relationships between the force exerted on a particle and the distance from the node of the standing wave in the acoustic field

distance from the node of the standing wave in the acoustic field. The acoustic force was calculated for a 10 μm upper sphere in ultrasound fields with $E_{ac} = 15$ (dashed line) and 30 J/m^3 (solid line) at a frequency of 500 kHz. The particle experiences an acoustic radiation force in the direction of the node of the standing wave. This force maximizes at a middle position between the node and the antinode. In this acoustic field, generated in water, the usual particles are aggregated at the node regardless of the E_{ac}, size, and acoustic properties of the particle. Figure 2.2 illustrates this principle.

A sedimentation force exerted on a particle in a medium is given by:

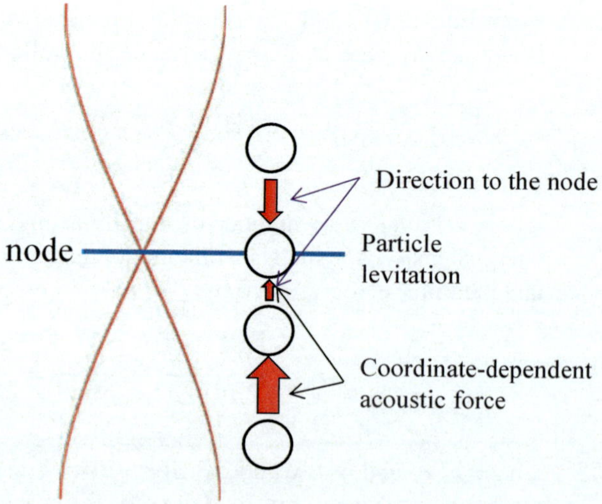

Fig. 2.2 Schematic of particle levitation in the acoustic field

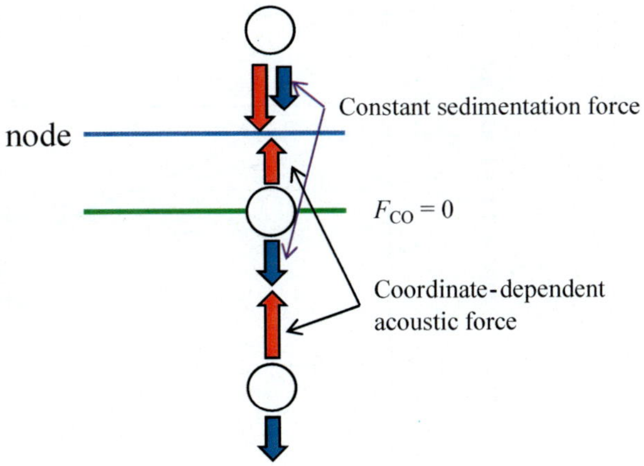

Fig. 2.3 Schematic of the particle aggregation in the CAG field

$$F_{sed} = \frac{4}{3}\pi r^3(\rho - \rho^*)g \tag{2.65}$$

where g is the gravitational acceleration. Thus, when the ultrasound standing wave is generated vertically, a CAG field is formed. In the CAG field, the particles simultaneously experience F_{ac} and F_{sed} (Fig. 2.3). The particles move along a resultant force gradient and then levitate at a position where the two forces are balanced.

The combined force F_{co} is given by:

$$F_{co} = F_{ac} + F_{sed} \tag{2.66}$$

Substituting Eqs. (2.63) and (2.65) into Eq. (2.66), we obtain the following equation:

$$F_{co} = \frac{4}{3}\pi r^3\{(\rho - \rho^*)g - kE_{ac}A\sin(2kz)\} \tag{2.67}$$

Figure 2.4 shows the relationships between the force exerted by a 10 μm copper particle and the distance from the node of the standing wave in the CAG field. The levitation coordinate of the particles is shifted downward from the node of the standing wave. In addition, the particles cannot be levitated at $E_{ac} = 15$ J/m^3.

The levitation coordinate of the particle was determined from $F_{co} = 0$. Thus, the following equation is obtained:

$$z = \frac{1}{2k}\sin^{-1}\left\{\frac{(\rho - \rho^*)g}{kAE_{ac}}\right\} \tag{2.68}$$

This equation indicates that z depends on the density and compressibility of the particle, but it is independent of the particle size.

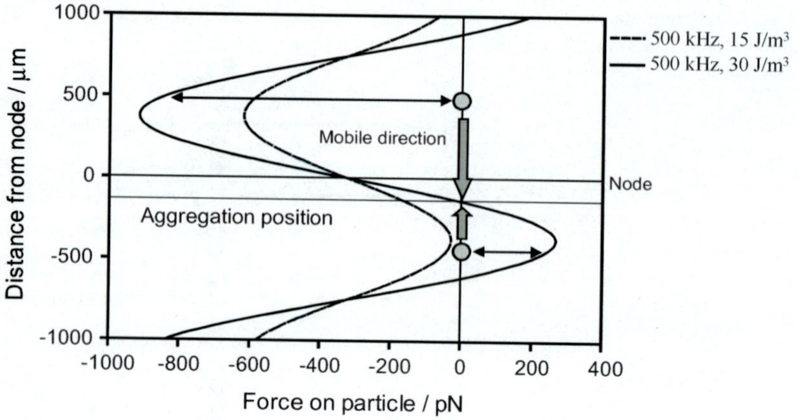

Fig. 2.4 Relationships between the force exerted on a particle and the distance from the node of the standing wave in the CAG field

In this study, an ultrasound standing wave is generated by applying an AC voltage to a piezoelectric ceramic element. The relationship between E_{ac} and voltage, V, is given by [5, 6]:

$$E_{\mathrm{ac}} = \alpha V^2 \tag{2.69}$$

where a is a device-dependent parameter. Substituting Eq. (2.69) into Eq. (2.68), we obtain:

$$z = \frac{1}{2k}\sin^{-1}\left\{\frac{(\rho - \rho^*)g}{kA\alpha V^2}\right\} \tag{2.70}$$

This indicates that the levitation coordinate of the particle is expressed as a function of the applied voltage.

The device-dependent parameter, a, can be determined by analyzing the V–z curve for particles with known density and compressibility based on Eq. (2.70).

References

1. Dinikov AA (1994) Acoustic radiation pressure on a compressible sphere in a viscous fluid. J Fluid Mech 267:1–22
2. Dinikov AA (1994) The crack tip region in hydraulic fracturing. Proc R Soc London Ser A 447:39–48
3. King LV (1934) On the acoustic radiation pressure on spheres. Proc R Soc London Ser A 147:212–240
4. King LV (1935) On the acoustic radiation pressure on circular discs: inertia and diffraction corrections. Proc R Soc London Ser A 153:1–16

5. Yasuda K, Umeura S, Takeda K (1995) Concentration and fractionation of small particles in liquid by ultrasound. Jpn J Appl Phys 34:2715–2720
6. Hatanaka S, Mitome H, Tuziuti T, Kozuka T, Kuwabara M, Asai S (1999) Relationship between a standing-wave field and a sonoluminescing field. Jpn J Appl Phys 38:3053–3057

Chapter 3
Principle of Detection Based on Particle Levitation in Coupled Acoustic-Gravitational Field

Abstract In this chapter, the detection principle based on particle levitation in a coupled acoustic-gravitational (CAG) field is theoretically described. A detection concept based on particle levitation is described in Sect. 3.1. The dependence of the levitation coordinate on the density and size of the particle is simulated in Sect. 3.2. Moreover, the effect of the thickness of a gold layer on the microparticle on the levitation coordinate is investigated in Sect. 3.3. In addition, the effect of gold nanoparticle (AuNP) binding on the levitation of a microparticle is discussed in Sect. 3.4. Because a device-dependent parameter is important for the evaluation of the actual binding number of AuNPs, its determination is described in Sect. 3.5.

Keywords Levitation coordinate shift · Density effect · Size effect · Gold nanoparticle binding

3.1 Basic Concept of Detection Based on Particle Levitation

As observed in Eq. (2.70), particles are levitated at different coordinates according to their density and compressibility in the CAG field [1–5]. A reaction on the surface of the particles induces changes in these acoustic properties. The density change of a microparticle can be caused by the binding of AuNPs, which lowers the levitation coordinate (z) of the microparticle in the CAG field. Thus, we can evaluate the density change from a shift of z (Δz) and eventually determine the number of target molecules that mediate the interparticle surface reaction.

Figure 3.1 shows the detection principle in the present study. This detection scheme requires high-affinity reactions for almost irreversible interparticle binding. The avidin–biotin reaction (dissociation constant, $K = 10^{-15}$ M) [6], antigen–antibody reaction ($K = 10^{-7}$–10^{-2} M) [7], and DNA hybridization ($K = 10^{-7}$–10^{-12} M) [8] can be utilized to realize the present concept. In this chapter, the acoustic properties of the microparticles are modified by the binding of AuNPs through the avidin–biotin reaction and DNA hybridization.

© The Author(s), under exclusive license to Springer Nature Singapore Pte Ltd. 2021
A. Miyagawa, *Acoustic Levitation-Based Trace-Level Biosensing*, Springer Theses,
https://doi.org/10.1007/978-981-16-1425-5_3

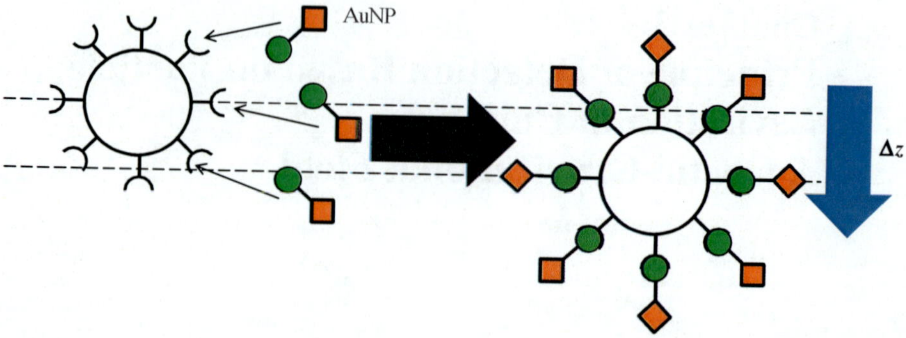

Fig. 3.1 Schematic of the levitation coordinate change by binding of AuNPs on a microparticle through a specific reaction in the CAG field

3.2 Levitation Coordinates of Microparticles with Different Densities and Sizes

The levitation coordinate of a gold-plated microparticle was calculated to discuss the impact of the density on the levitation coordinate of a microparticle. For a polymer microparticle with a thin gold layer with a uniform thickness, l, on the surface, the relationship between its density and the levitation coordinate was simulated by varying the radius of a microparticle, r. The density of the gold-plated microparticles (ρ') is represented by:

$$\rho' = \frac{\frac{4}{3}\pi r^3 \rho_{PM} + \frac{4}{3}\pi\left((r+l)^3 - r^3\right)\rho_{Au}}{\frac{4}{3}\pi(r+l)^3} \tag{3.1}$$

where $\rho_{PM}(\rho_{PM} = 1.052\ \mathrm{g\ cm^{-3}})$and ρ_{Au} ($= 19.32\ \mathrm{g\ cm^{-3}}$) are the densities of a PS microparticle and that of the gold layer, respectively [9]. For $l = 15$ nm, the relationship between r and ρ' can be calculated from Eq. (3.1), as shown in Fig. 3.2. As r increases, ρ' becomes smaller. Therefore, the use of a smaller particle allows higher detection sensitivity because a large density change is induced.

The compressibility of the gold-plated microparticles (γ') is given by the following equation:

$$\gamma' = \phi_{PM}\gamma_{PM} + \phi_{Au}\gamma_{Au} \tag{3.2}$$

where ϕ_{PM} and ϕ_{Au} are the volume ratios for the PS and gold in the microparticle, respectively, and γ_{PM} and γ_{Au} are the compressibility of the PS particle ($2.3753 \times 10^{-10}\ \mathrm{Pa^{-1}}$) and gold ($5.88 \times 10^{-12}\ \mathrm{Pa^{-1}}$) [10], respectively. Figure 3.3 shows the relationship between the calculated γ' and r. The change in γ' is marginal in the range of $r = 5$–$20\ \mu$m ($\Delta\gamma = 1.58 \times 10^{-12}\ \mathrm{Pa^{-1}}$). Thus, because γ_{Au} is smaller than

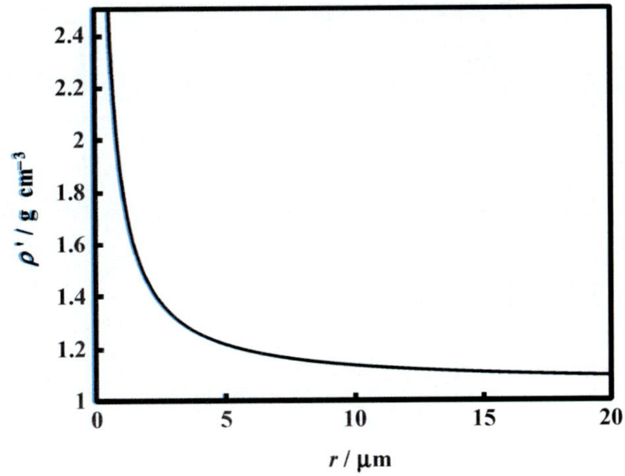

Fig. 3.2 Relationship between r and ρ' for the gold-plated PS microparticle with $d = 15$ nm

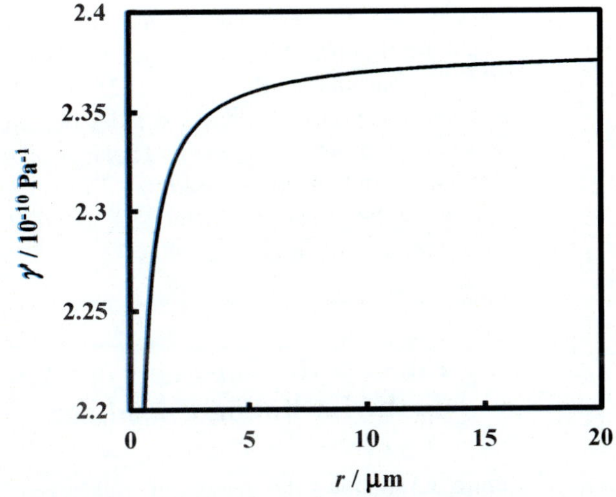

Fig. 3.3 Relationship between r and γ' for the gold-plated PS microparticle with $l = 15$ nm

γ_{PM} by two orders of magnitude, the impact of γ' on z is also negligible. Thus, z is mainly determined by the density change.

The dependence of z on r of the PS particle with a gold layer ($l = 15$ nm) was calculated at a varying voltage (V). Substituting ρ' into Eq. (2.70) allows the calculation of z at various V, as shown in Fig. 3.4. As V decreases, z becomes smaller because the acoustic radiation force F_{ac} also decreases. In addition, z decreases with decreasing r, and the slope of the z–r curve becomes larger as r decreases because the density becomes larger for smaller r. It is desirable for highly sensitive analysis that a minute amount of the density change can be detected as a large change in z. A smaller microparticle is suitable from this perspective. However, F_{ac} decreases as the particle size decreases because F_{ac} is a function of r^3; a small particle may not be levitated or it may take a long time before the particle reaches the equilibrium position. For the

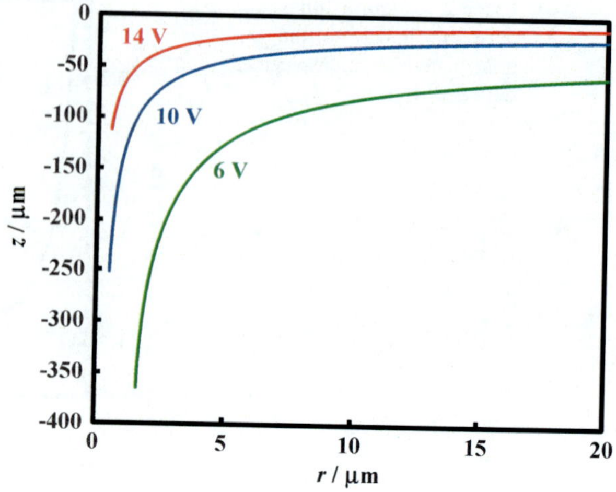

Fig. 3.4 Changes in z with r at $V = 6$, 10, and 14 V. $a = 0.0057$

particles used in this work, size distribution of microparticles is: epoxide particles with 4.4 ± 2.3 µm, PS particles with 10.4 ± 0.5 µm, and polymethyl methacrylate (PMMA) particles with 9.57 ± 0.21 µm. The effect of the size variation of the microparticles on ρ' becomes smaller as r increases because the slope of the r-ρ' plots shown in Fig. 3.2 is smaller for larger r. A larger particle is desirable from these viewpoints. In contrast, it becomes difficult to determine z with high precision for excessively large particles. Thus, we concluded that microparticles with r ~ 5 µm are suitable for the present system.

3.3 Effect of the Thickness of a Gold Layer on the Levitation Coordinate of a Microparticle

Figure 3.5 a shows the density change *versus* l of a microparticle with $r = 5$ µm. In a range of $l = 0$–100 nm, ρ' is proportional to l. Substituting the resulting ρ' into Eq. (2.70) allowed the calculation of z as shown in Fig. 3.5 B, in which the results simulated at $V = 6$, 10, and 14 V are depicted by red, blue, and green curves, respectively. The slope of l-z plots becomes more negative as V decreases. The change in z increases by lowering V, indicating that a smaller density change can be detected at lower V.

Figure 3.6 shows an enlargement of Fig. 3.5 in the range of $l = 0$–2 nm. Assuming that the standard deviation of z of the microparticle measured in the CAG field is 10 µm, the theoretical detection limit of the density change is 0.012 and 0.016 g cm^{-3} at $V = 5$ and 6 V, respectively.

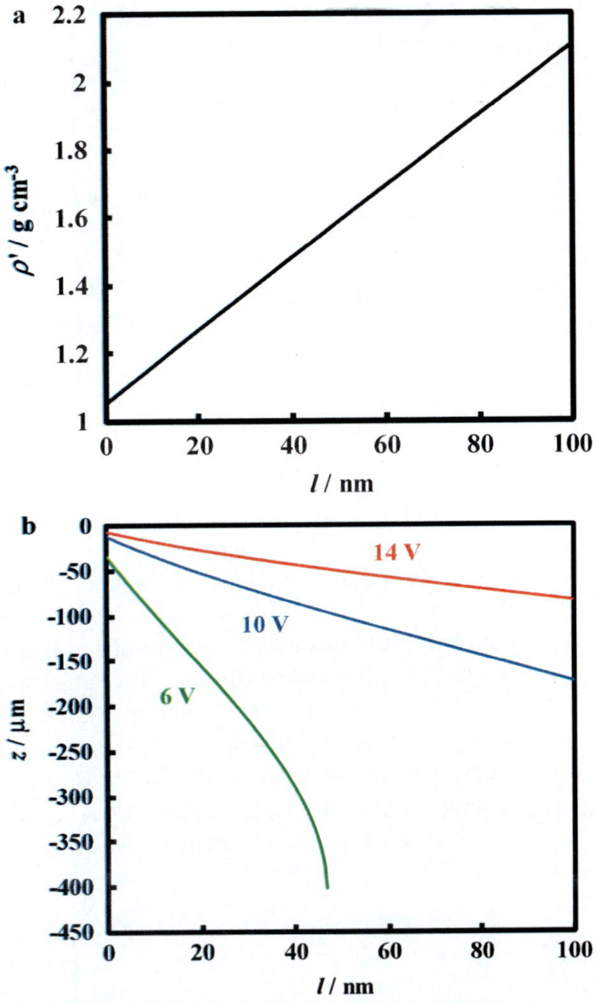

Fig. 3.5 **a** Relationship between l and ρ' for the gold-plated polystylene microparticle with $r = 5$ μm and **b** change in z with l at $V = 6$, 10, and 14 V. $\alpha = 0.0057$

3.4 Effect of AuNP Binding on the Levitation Coordinate of a Microparticle

The density of AuNP-bound microparticles (ρ') was calculated from the volume ratio of AuNPs to a polymer microparticle. ρ' is given by the following equation:

$$\rho' = \frac{\frac{4}{3}\pi {r_{\mathrm{PM}}}^{3}\rho_{\mathrm{PM}} + \frac{4}{3}\pi {r_{\mathrm{AuNP}}}^{3}\rho_{\mathrm{AuNP}} n_{\mathrm{AuNP/MP}}}{\frac{4}{3}\pi \left({r_{\mathrm{PM}}}^{3} + {r_{\mathrm{AuNP}}}^{3}\right)} \tag{3.3}$$

where $\boldsymbol{\rho}_{\mathrm{AuNP}}$ (=19.32 g cm^{-3}) is the density of AuNPs, $n_{\mathrm{AuNP/MP}}$ is the number ratio of AuNPs to microparticles, and r_{PM} and r_{AuNP} are the radii of the microparticles

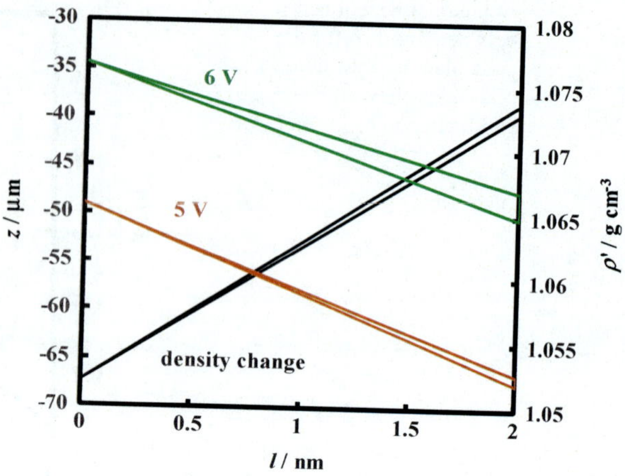

Fig. 3.6 Enlargement of combined plots A and B in Fig. 3.5

and AuNP, respectively. Assuming $n_{\mathrm{AuNP/MP}} = 1500$ and $r_{\mathrm{AuNP}} = 50$ nm, $\boldsymbol{\rho}$ as a function of r_{PM} was calculated, as shown in Fig. 3.7. As r_{PM} decreases, ρ' increases, indicating that a smaller particle is advantageous for high sensitivity in the present detection scheme. The advantages and disadvantages of using smaller particles have already been discussed for gold-plated microparticles. In this study, microparticles with $r_{\mathrm{PM}} \sim 5$ µm were used for the same reason.

For $r_{\mathrm{PM}} = 5$ µm, the levitation coordinate of an AuNP-bound microparticle was calculated by substituting Eq. (3.3) into Eq. (2.70). Figure 3.8 shows the relationship between Δz and $n_{\mathrm{AuNP/MP}}$ at $V = 5$ V. Because z is a complex function of ρ', as shown in Eq. (2.70), z is not proportional to $n_{\mathrm{AuNP/MP}}$, particularly for larger $n_{\mathrm{AuNP/MP}}$ values. However, a linear relationship between z and $n_{\mathrm{AuNP/MP}}$ was confirmed in the range of

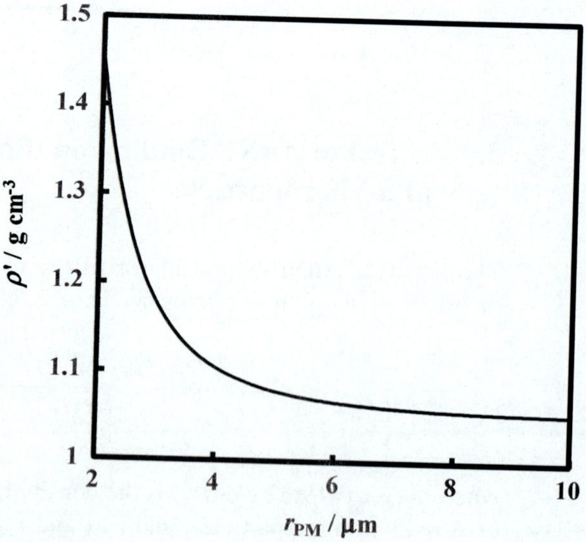

Fig. 3.7 Relationship between r_{PM} and ρ' for AuNP-bound PS microparticles

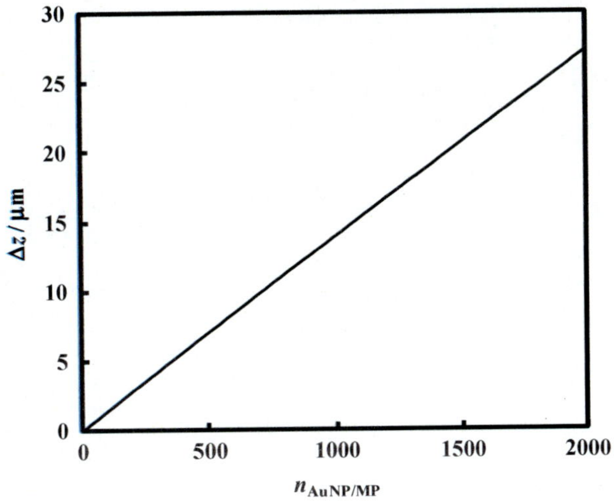

Fig. 3.8 Relationship between $n_{AuNP/MP}$ and Δz at $V = 5$ V for AuNP-bound microparticles

$n_{AuNP/MP} = 0$–15,000. This indicates that for a measured $\Delta z = 20$ μm caused by the binding of AuNPs, 1500 AuNPs can be detected, which means that 1500 molecules are involved in an interparticle reaction.

3.5 Determination of Device-Dependent Parameter

The device-dependent parameter, a, should be determined to evaluate the actual number of AuNPs bound to a microparticle from Δz. PMMA particles with densities of 1.265 and 1.188 g cm^{-3} were used to study the effect of the density on the levitation coordinate. Figure 3.9 shows the relationships between V and z for the PMMA particles with densities of 1.188 (black) and 1.265 g cm^{-3} (red). As discussed above, z decreases as V decreases because F_{ac} becomes weak. The threshold voltage (V_{th}), which is designed as the highest voltage incapable of levitating a particle, was 5.6 and 6.3 V for 1.188 and 1.265 g cm^{-3} PMMA particles, respectively. The solid curves in Fig. 3.9 represent the results of curve-fitting based on Eq. (2.70) with the device-dependent parameter, a, assumed as the fitting parameter [11]. The same compressibility for both PMMAs ($\gamma = 1.54 \times 10^{-10}$ Pa^{-1}) was assumed. The a values for PMMAs with densities of 1.188 and 1.265 g cm^{-3} were determined to be 0.0408 and 0.0388, respectively. Because these values were almost identical, the average value, that is, $a = 0.040$, was used for the determination of the number of AuNPs bound to PMMA microparticles, as discussed in Chapter 6. It should be noted that this value is not applicable to all cases because a is changed daily.

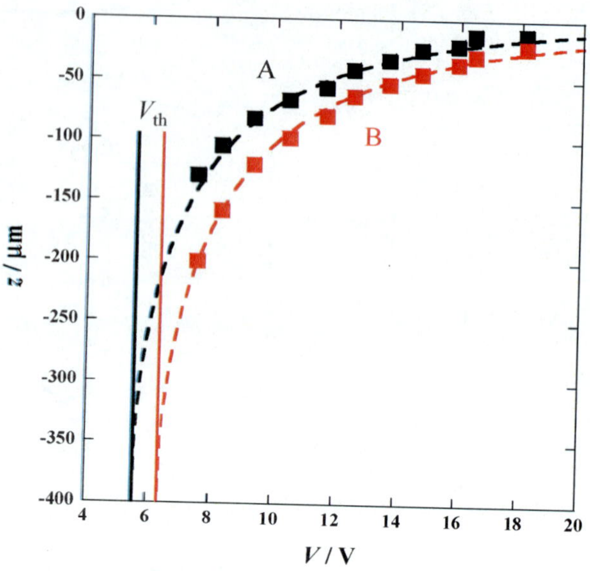

Fig. 3.9 Voltage-dependent levitation coordinate of PMMA microparticles with different densities; black: 1.188 and red: 1.265 g cm^{-3}. $a = 0.0408$ and 0.0388 for PMMAs with 1.188 and 1.265 g cm^{-3}, respectively

References

1. Masudo T, Okada T (2001) Particle characterization and separation by a coupled acoustic-gravity field. Anal Chem 73:3467–3471
2. Masudo T, Okada T (2001) Ultrasonic radiation—novel principle for microparticle separation. Anal Sci 17:i1341–i1344
3. Masudo T, Okada T (2006) Particle separation with ultrasound radiation force. Curr Anal Chem 2:213–227
4. Masudo T, Okada T (2004) Elution control of microparticles with a coupled acoustic-gravity field and orthogonal laminar flow. Anal Sci 20:753–755
5. Kanazaki T, Hirawa S, Harada M, Okada T (2010) Coupled acoustic-gravity field for dynamic evaluation of ion exchange with a single resin bead. Anal Chem 82:4472–4478
6. Green NM (1963) Avidin 3. The Nature of the biotin-binding Site. Biochem J 89:599–609
7. Lin A, Lee ASY, Lin CC, Lee CK (2006) Determination of binding constant and stoichiometry for antibody-antigen interaction with surface plasmon resonance. Curr Proteomics 3:271–282
8. Stevens PW, Henry MR, Kelso DM (1999) DNA hybridization on microparticles: determining capture-probe density and equilibrium dissociation constants. Nucleic Acid Res 27:1719–1727
9. Kagakubinran (Chemical Index) (2004). Maruzen, Tokyo
10. Takei H, Okamoto T (2016) Morphology effects of cap-shaped silver nanoparticle films as a SERS platform. Anal Sci 32:287–293
11. Miyagawa A, Okada Y, Okada T (2020) Aptamer-based sensing of small organic molecules by measuring levitation coordinate of single microsphere in combined acoustic-gravitational field. ACS Omega 5:3542–3549

Chapter 4
Detection of the Avidin–Biotin Reaction

Abstract In this section, a detection scheme based on the levitation coordinate shift of a microparticle by the binding of AuNPs is demonstrated. The relationship between the particle density and the levitation coordinate is studied using gold-plated PMMA microparticles in Sect. 4.2. In addition, the levitation coordinate change of epoxy microparticles (EPs) by AuNP-binding is discussed in Sect. 4.3. A zmol sensing scheme based on the levitation of a single PS microparticle is shown in Sect. 4.4. Finally, this method was applied to the quantification of dissolved biotin in Sect. 4.5.

Keywords Epoxy particle · Gold-plated particle · Avidin–biotin reaction · Zmol detection · PM detection

4.1 Experimental

4.1.1 Preparation of Epoxy Microparticles

EPs were prepared by a membrane emulsification process, [1] as shown in Fig. 4.1. Ethylene glycol dimethacrylate and glycidil methacrylate were mixed at a volume ratio of 4:1. To the solution, 2,2-azobis (isobuthyronitrile) (1w/v%) was added as a polymerization initiator. The resultant solution was pushed into an aqueous 0.2% methyl cellulose solution through a porous silica membrane with a pore size of 3 μm (DC03N, SPG Technology Co., Ltd). The aqueous solution was stirred at 400 rpm using a magnetic stirrer. Therefore, microdroplets formed on the membrane were dispersed in the aqueous phase by shear flow. The flow rate of the organic phase through the membrane was maintained at 0.025 mL min^{-1} using a syringe pump (Harverd Apparatus, model Pump 11). The emulsions were kept at 78 °C for 4 h for polymerization. The EPs were collected by centrifugation and washed with deionized water and ethanol. The size distribution of the EPs (546 particles) was measured by microscopic observations. The result is shown in Fig. 4.2. The average diameter was determined to be 4.4 μm.

The EP density was determined by heavy liquid separation using sodium polytungstate solutions with various densities, which were calibrated using gravimetry. The density of the EPs was determined to be 1.304 g cm^{-3}.

© The Author(s), under exclusive license to Springer Nature Singapore Pte Ltd. 2021

A. Miyagawa, *Acoustic Levitation-Based Trace-Level Biosensing*, Springer Theses,
https://doi.org/10.1007/978-981-16-1425-5_4

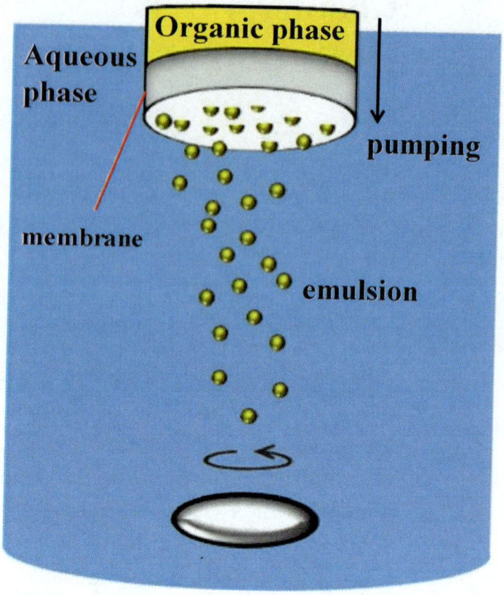

Fig. 4.1 Schematic of the membrane emulsification method used for EP preparation

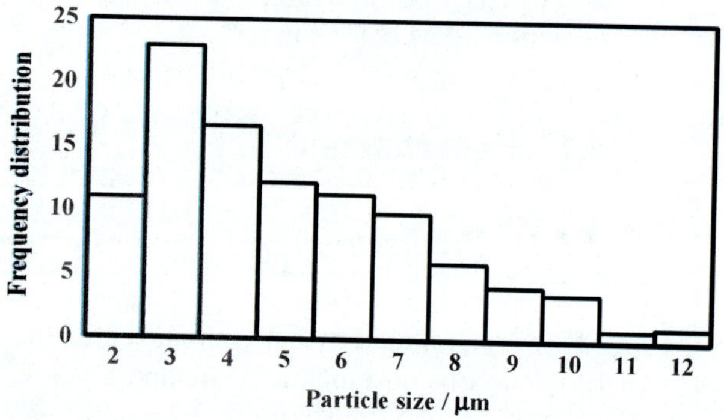

Fig. 4.2 Size distributions for EPs. The diameters of 546 particles were measured

Avidin-modified EPs were prepared by the reaction of epoxide groups on the EP surface with the amino group of the lysine or arginine residue in the avidin molecule. The maximum capacity of avidin on EP was calculated from the surface area of EP and the area of avidin (22.0 nm^2) [2]. The reaction of epoxide groups with amino groups usually proceeds under basic conditions. Therefore, EPs were added to a borate buffer (pH 9.18) containing an excess of avidin molecules (10 times the maximum capacity). The reaction time was 5 h. The avidin-modified EPs were washed with deionized water and dried under reduced pressure. The number of avidin molecules on a single EP was determined to be 9.44×10^3 using the reaction with biotin-4-fluorescein [3, 4].

AuNP-bound EPs were prepared by the avidin–biotin reaction. The avidin-modified EPs were mixed with biotin-modified AuNPs with a diameter of 50 or 100 nm (Cytodiagnostics, Burlington, Canada).

4.1.2 *Preparation of AuNP-Bound Polystyrene Microparticles*

Avidin-modified PS microparticles were purchased from Micromod (Rostock, Germany). The sizes of the 343 avidin-modified PS particles were measured using a microscope with a resolution of 1 μm, and the size distribution is shown in Fig. 4.3. The particle size of PS was 10.4 ± 0.5 μm. The number of avidin molecules on a PS particle was determined to be 1.09×10^6 by the method described above. The concentration of the PS particles in the stock solution was determined to be 5.50×10^7 mL^{-1} by counting the number of particles contained in 2 μL aliquots of the suspension using a microscope. The concentration and diameter of AuNPs were measured using the spectroscopic method reported by Fernig and co-workers [5]. The relationship between the diameter of AuNP (d) and its SPR peak (λ_{spr}) is represented by:

$$d = \frac{\ln\left(\frac{\lambda_{spr}-\lambda_0}{L_1}\right)}{L2} \tag{4.1}$$

where λ_0 (=512 nm), L_1 (=6.53), and L_2 (=0.0216) are parameters calculated from theoretical values. Figure 4.4 shows the absorption spectrum of the 80-fold dilution of the original AuNP suspension. The AuNP diameter, d, was determined to be 101 nm from $\lambda_{spr} = 570$ nm. The number of AuNPs (N) is given by the following equation:

$$N = \frac{A_{450} \times 10^{14}}{d^2\left[-0.295 + 1.36\exp\left(-\left(\frac{d-96.8}{78.2}\right)^2\right)\right]} \tag{4.2}$$

where A_{450} is the absorbance of the AuNP suspension at 450 nm. Thus, N was determined to be 1.76×10^{11} mL^{-1}.

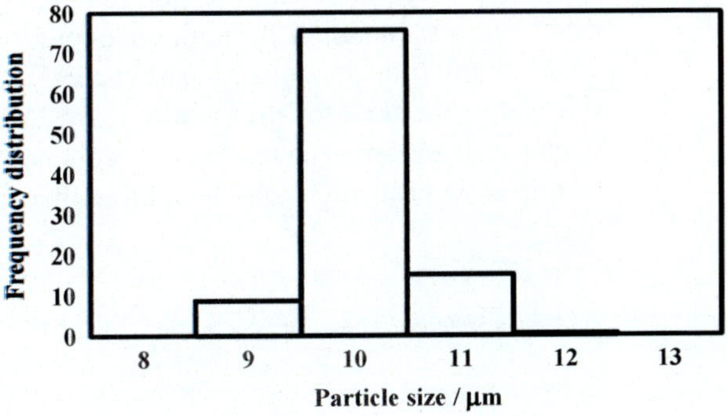

Fig. 4.3 Size distribution for PS microparticles. The diameters of 343 particles were measured

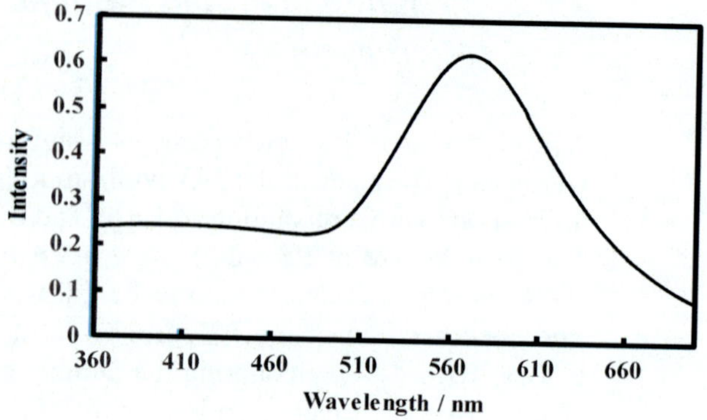

Fig. 4.4 Absorption spectra of AuNP suspension. The AuNP stock solution underwent 80-fold dilution

The avidin-modified PS microparticles were mixed with the biotin-modified AuNPs for 4 h. The binding number of AuNPs per PS microparticle was controlled by changing the number ratio of AuNPs to PS microparticle ($r_{AuNP/PS}$) added to the reaction mixture.

4.1.3 Experimental Setup

Figure 4.5 shows the experimental setup used in the present study. The sinusoidal signal, which was generated by a function generator (TEXIO, model FGX-2200), was amplified by a bipolar high-speed amplifier (NF Electric, model 4015). The amplified signal was applied to a transducer (2 × 2 cm lead zirconate, resonance frequency of 500 kHz, Fuji Caramics) placed on a three-dimensional stage, which allowed the adjustment of the vertical and horizontal positions. The levitation of microparticles was observed by a charge-coupled device camera (Olympus, model CS220) through a zoom lens (maximum magnification of × 24). The levitation coordinate was defined as the middle of the upper and lower edge positions of a single particle or particle aggregate.

A fused silica glass cell (30 mm length, 8 mm width, and 12.62 mm thickness), which had a rectangular through-channel (3.0 mm width and 1.5 mm height), was pasted on the transducer using nail enamel as an adhesive. When the cell was filled with water, the node of the standing wave (500 kHz) was formed at the center of the channel. Because the resonance frequency was changed daily, the frequency was optimized by ensuring stable levitation of particles before measurements.

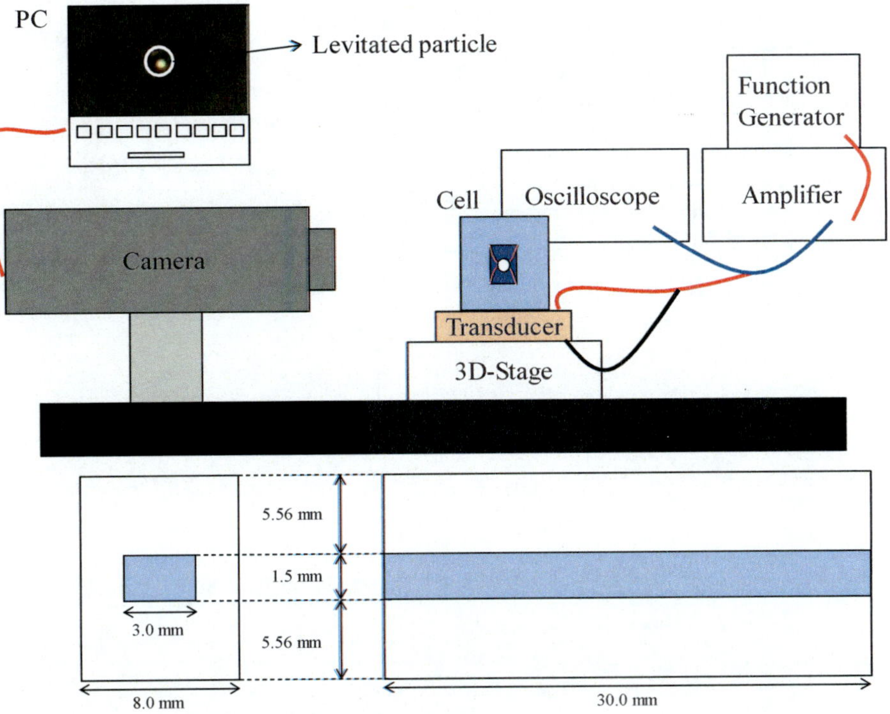

Fig. 4.5 Schematic of the experimental setup used in the present study

4.2 Levitation Behavior of Gold-Plated Polymethyl Methacrylate Particles

The levitation behavior of the gold-plated PMMA microparticles with various thicknesses of gold layer (l) was evaluated in the CAG field. Figure 4.6 shows the relationships between V and z of gold-plated PMMA particles with various l values. With decreasing V, z decreased because of a decrease in F_{ac}. When V became lower than a threshold value, the F_{ac} was not sufficiently large to keep the particles levitated. The levitation coordinate became smaller as l increased, indicating that z depends on the density of the microparticles. These behaviors agree with the predictions from Fig. 3.4b in Sect. 3.3. The densities of the PMMA and gold-plated PMMA particles, which were determined using the heavy liquid method described in Sect. 4.1, are listed in Table 4.1. The densities estimated from Eq. (3.1) are also shown in this table. The measured densities of gold-plated PMMA almost agree with the corresponding values calculated using Eq. (3.1). The curves in Fig. 4.6 represent the results of curve fitting based on Eq. (2.70) with compressibility as a fitting parameter for each curve, and adequately reproduce the experimental values. Table 4.2 lists the obtained γ for gold-plated PMMAs with various l values. Although the γ values of gold-plated PMMAs are different from that of the bare PMMA, no significant

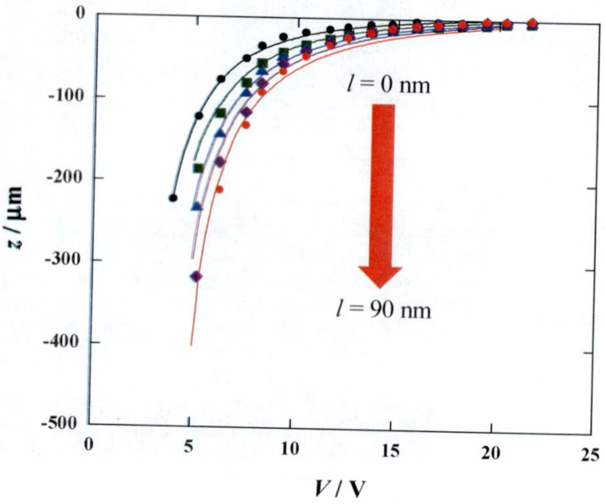

Fig. 4.6 Relationships between V and z of gold-plated PMMA particles with various l values (0, 30, 50, 70, and 90 nm). Curve-fitting with $a = 0.232$ and γ as fitting parameters. The γ values are listed in Table 4.2

Table 4.1 Densities of PMMA and gold-plated PMMA particles

l/nm	Measured density/g cm^{-3}	Calculated density/g cm^{-3}
0	1.265	–
30	1.560	1.59
50	1.775	1.80
70	1.948	2.00
90	2.160	2.21

Table 4.2 Compressibility of bare and AuNP-plated PMMA particles determined by curve-fitting in Fig. 4.6

l/nm	Compressibility/10^{-10} Pa^{-1}
0	4.30
30	2.81
50	2.69
70	2.91
90	3.07

changes in γ with l were found. This indicates that the CAG field may strongly recognize the material of the particle surface. In the case of gold-plated PMMA, the surface of the PMMA particle is completely covered by the gold layer, which leads to a change in γ. For the AuNP-bound microparticles, AuNP cannot completely cover the microparticles, suggesting that AuNP-binding to microparticles does not influence γ. Thus, compressibility is not a very important parameter when the levitation of a microparticle is compared, as the microparticle surface is not significantly modified. Hence, the density of the gold-plated microparticles can be determined by measuring their levitation coordinates.

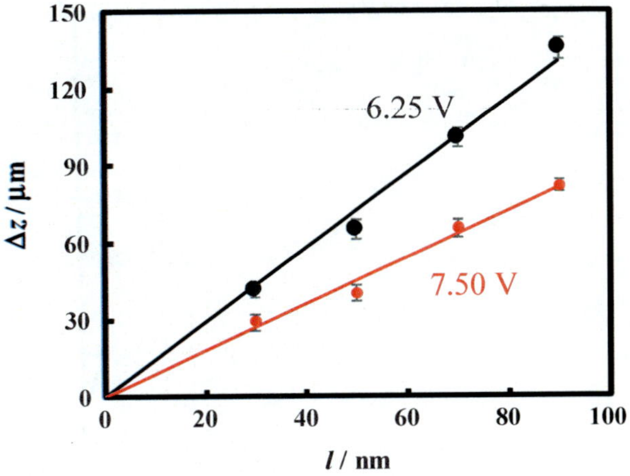

Fig. 4.7 Relationships between Δz of gold-plated PMMAs and l at $V = 6.25$ (black) and 7.50 V (red)

The difference in z between the PMMA and gold-plated PMMA particles (Δz) is, hereafter, used for the discussion of changes in the density of a microparticle. Figure 4.7 shows the relationships between Δz and l at $V = 6.25$ and 7.50 V. Evidently, Δz is proportional to l at a given voltage. The detection limit was defined as three times the standard deviation of z (3σ). The detection limit of l was 6.3 nm for $V =$ 6.25 V and 10.0 nm for $V = 7.50$ V. These results suggest that a small increase in the density, which is induced by AuNP binding to microparticles, can be detected as Δz.

4.3 Levitation Behavior of Gold Nanoparticle-Bound Epoxy Particles

The effect of AuNP binding on the levitation coordinate of microparticles is schematically shown in Fig. 3.1. Here, AuNPs were bound to EPs through the avidin–biotin reaction, and the effect of AuNP-binding to EPs on z was studied. Multiple AuNP-bound EPs were introduced into the cell, and the levitation behavior of an EP aggregate was analyzed. Figure 4.8 shows the V–z plots of EPs, 50 nm AuNP-bound EPs ($r_{\mathrm{AuNP/EP}} = 38{,}260$), and 100 nm AuNP-bound EPs ($r_{\mathrm{AuNP/EP}} = 8350$). Evidently, 100 nm AuNP-bound EPs were levitated at coordinates lower than 50 nm AuNP-bound EPs at a given V. Curve-fitting of V-z plots was performed to determine the acoustic parameters of these particles. The densities of 50 nm and 100 nm AuNP-bound EPs were determined to be 1.416 and 1.518 g cm^{-3}, respectively. These density changes occur if 4000 50 nm AuNPs and 1000 100 nm AuNPs bind to a single EP. The binding number of AuNPs calculated from z-V plots was approximately one-ninth that of $r_{\mathrm{AuNP/EP}}$ for both 50 nm and 100 nm AuNP-bound EPs. Interestingly, the coverage of AuNPs on the EP was calculated to be 13% for both AuNP-bound EPs. Thus, the large density change for the 100 nm AuNP-bound EP is attributed to

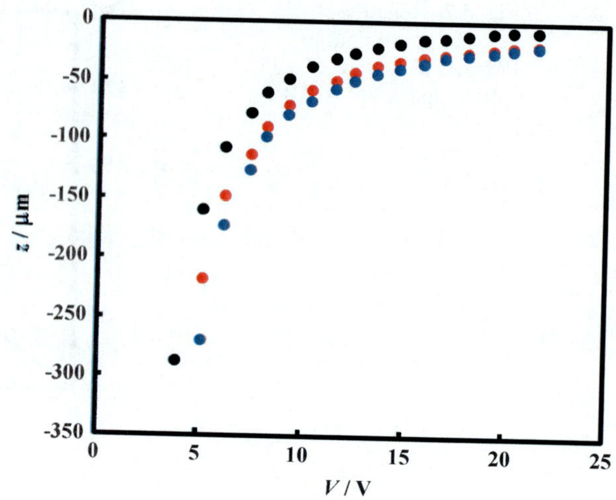

Fig. 4.8 Voltage dependence of z for EPs (black), 50 nm AuNP-bound EPs (red), and 100 nm AuNP-bound EPs (blue)

the large diameter of AuNPs, as the coverage was the same for both AuNP-bound EPs. The coverage of AuNPs under the above conditions does not increase because $r_{AuNP/EP}$ is already sufficiently large to cover the entire EP surface with AuNPs if all of the added AuNPs are bound. This may be caused by the steric hindrance between AuNPs and by the electrostatic repulsion between the AuNPs bound on EP and free AuNPs. Hence, the larger AuNPs induce larger Δz and enable the detection of the interparticle reaction with high sensitivity. In subsequent experiments, 100 nm AuNPs were used.

The levitation of EPs treated with various amounts of AuNPs ($r_{AuNP/EP}$) was studied. Figure 4.9 shows images of the levitation of EPs and AuNP-bound EPs with $r_{AuNP/EP} = 321$ and 642 at $V = 5.0$ V. The levitation coordinate of the EPs was lowered as $r_{AuNP/EP}$ increased. Figure 4.10 shows the relationship between Δz and $r_{AuNP/EP}$ at $V = 5.0$ V. In the range of $r_{AuNP/EP} = 0 - 642$, Δz was proportional to $r_{AuNP/EP}$. For larger $r_{AuNP/EP}$, the increase in Δz became smaller, suggesting that the further binding of AuNPs on EP was hindered for the same reason mentioned above. The standard deviation of z for EPs (σ) was 3 µm. When the detection limit was defined as 3σ, 110 AuNPs per EP were detectable in this system, indicating that 110 avidin–biotin reactions occurred on the surface of the microparticles. However, this value is the average calculated from a number of AuNPs and EPs because multiple EPs were used for measurements. Thus, a new scheme based on the levitation of a single microparticle is necessary for detecting an event occurring on the surface of a single particle.

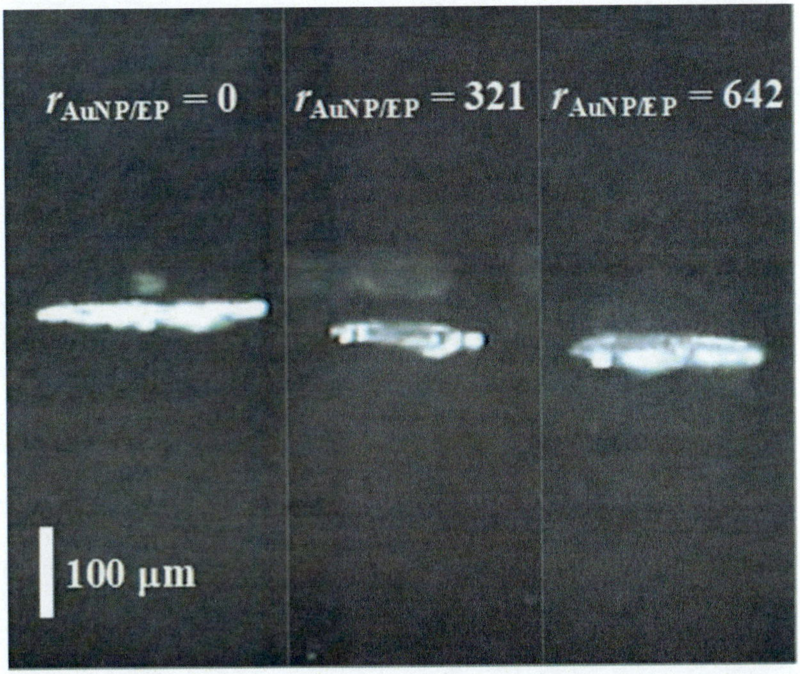

Fig. 4.9 Images of the levitation of EPs and AuNP-bound EPs with $r_{AuNP/EP} = 321$ and 642 at $V = 5.0$ V

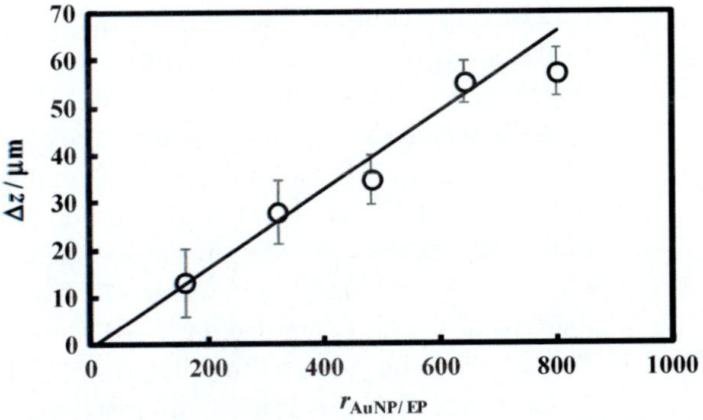

Fig. 4.10 Relationship between Δz of AuNP-bound EP and $r_{AuNP/EP}$ at $V = 5.0$ V

4.4 Measurements of Single Microparticle Levitation

Multiple EPs were used to establish the detection concept based on the measurement of a levitation coordinate shift. However, it was difficult to obtain measurable quantities from the behavior of a single EP particle because the size deviation of the EPs was large (±2.2 μm for the average diameter of 4.4 μm). PS microparticles were used because of the smaller size deviation. The response of a PS particle to an

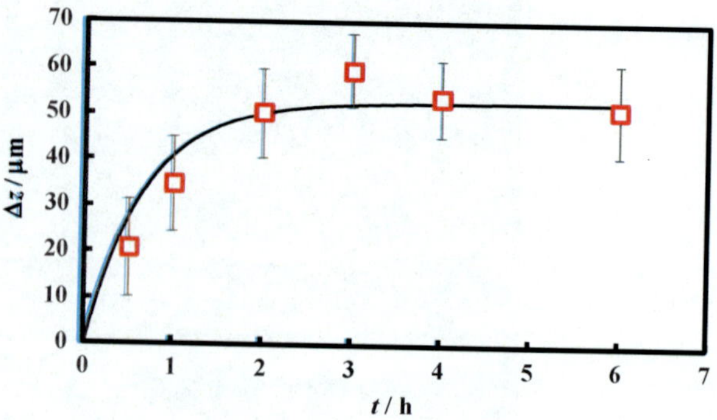

Fig. 4.11 Time dependence of Δz for the interparticle avidin–biotin reaction between PS microparticle and AuNPs. $r_{\mathrm{AuNP/PS}} = 4000$

acoustic force in the CAG field is slow because the density of PS (1.052 g cm^{-3}) is similar to that of water (0.997 g cm^{-3}), and, thus, F_{ac} is small. The PS microparticle typically reached equilibrium within 5 min. The time required for analysis should be reduced for practical purposes. In this study, the analysis time was not further considered because the main purpose was to establish the concept for detection based on a single particle behavior.

To evaluate the levitation coordinate of a single microparticle, a diluted PS microparticle suspension was introduced to the cell such that approximately 100 PS microparticles existed in the cell. This particle density was appropriate because no particle aggregation occurred, but one particle was often found in the microscopic view. A single PS microparticle was entrapped by the acoustic force in the observed region.

The time required for PS-AuNP interparticle binding by the avidin–biotin reaction was optimized by evaluating the time dependence of Δz. Figure 4.11 shows the dependence of Δz on the reaction time when a single PS microparticle was treated with 4000 AuNPs ($r_{\mathrm{AuNP/PS}} = 4000$). The change in the levitation coordinate became constant within 2–3 h after the start of the reaction. Therefore, the reaction time was maintained at 4 h for all experiments. The diffusion coefficients of the PS microparticles and AuNPs calculated from the Stokes–Einstein equation were 5×10^{-14} m^2 s^{-1} and 5×10^{-12} m^2 s^{-1}, respectively. These values are considerably smaller than the typical values for molecules (~10^{-10} m^2 s^{-1}). This suggests that the binding of PS particles with AuNPs through the avidin–biotin reaction takes a longer time than the molecular-based avidin–biotin reaction.

The reaction between avidin and biotin molecules can be represented by:

$$A + B \underset{k_{-1}}{\overset{k_1}{\rightleftarrows}} AB \tag{4.3}$$

where A and B are the binding sites in avidin and biotin molecules, respectively, and k_1 and k_{-1} are reaction rate constants. The rate equation for Eq. (4.3) is given by:

$$\frac{\mathrm{d}x}{\mathrm{d}t} = k_1([\mathrm{A}]_0 - x)([\mathrm{B}]_0 - x) - k_{-1}x \tag{4.4}$$

where $[\mathrm{A}]_0$ and $[\mathrm{B}]_0$ are the initial concentrations of the binding sites of avidin and biotin, respectively, and x is the concentration of the AB complex. Solving Eq. (4.4), we can obtain the time-dependent x represented by the following equations:

$$x = \beta + \frac{\alpha}{\delta e^{\alpha t}} \tag{4.5}$$

$$\alpha = \gamma - 2k_1\beta \tag{4.6}$$

$$\beta = \frac{\gamma + \sqrt{\gamma^2 - 4k_1^2[\mathrm{A}]_0[\mathrm{B}]_0}}{2k_1} \tag{4.7}$$

$$\gamma = k_1([\mathrm{A}]_0 + [\mathrm{B}]_0) + k_{-1} \tag{4.8}$$

$$\delta = -\frac{\alpha}{\beta} - k_1 \tag{4.9}$$

Thus, the rate constants can be determined by curve-fitting using Eqs. (4.5)–(4.9). The curve in Fig. 4.11, which was calculated using Eqs. (4.5)–(4.9) with $k_1 = 4.0 \times 10^5\ \mathrm{M}^{-1}\ \mathrm{s}^{-1}$ and $k_{-1} = 1.0 \times 10^{-4}\ \mathrm{s}^{-1}$, adequately reproduces the experimental values.

Figure 4.12 shows images of a single AuNP-bound PS microparticle in the CAG field. The levitation coordinate of the single PS particle, which was successfully entrapped, becomes lower as the number of AuNPs bound to the PS particle increases. Figure 4.13 shows the relationship between Δz and $r_{\mathrm{AuNP/PS}}$ at $V = 3.3$ V. This figure indicates that Δz is proportional to $r_{\mathrm{AuNP/PS}}$ in the range of $r_{\mathrm{AuNP/PS}} = 0$–5000 and becomes constant for a further increase in $r_{\mathrm{AuNP/PS}}$, suggesting that the interparticle reaction does not proceed at $r_{\mathrm{AuNP/PS}} > 5000$. The surface coverage of the PS particle for $r_{\mathrm{AuNP/PS}} = 5000$ is 12.5%; this coverage is almost the same as that determined for EPs. No additional AuNPs bind to the PS particle because of steric hindrance and electrostatic repulsion. The detection limit was 700 AuNPs per PS particle, which means that the formation of 700 avidin–biotin complexes was detected on a single PS particle. This value is higher than that obtained for EPs. This is mainly due to the smaller size of an EP. However, a diameter of ~ 10 μm is optimum considering the effect of size variation, as discussed in Sect. 3.4. Thus, the detection scheme for the interparticle reaction based on the levitation coordinate of a single microparticle is successfully demonstrated.

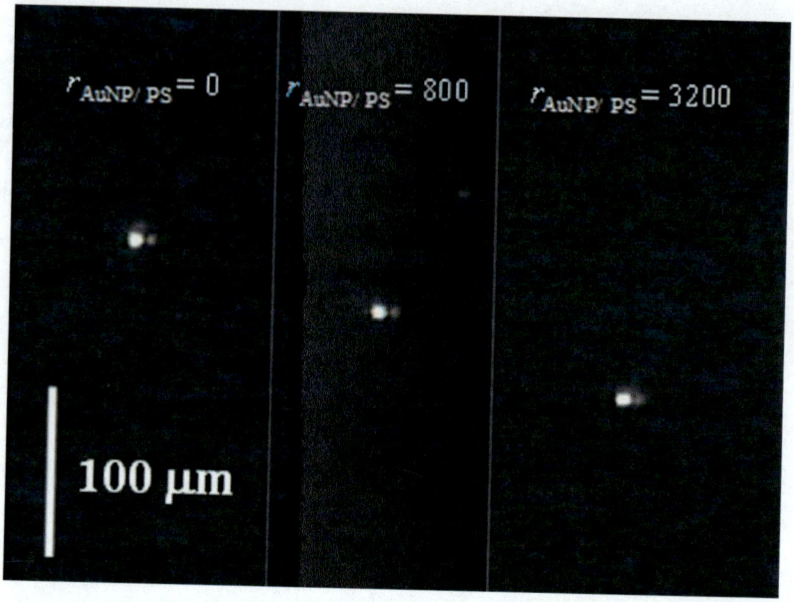

Fig. 4.12 Images of a single AuNP-bound PS microparticle levitated in the CAG field. $r_{AuNP/PS} = 0$, 800, and 3200, respectively

Fig. 4.13 Relationship between Δz of AuNP-bound PS and $r_{AuNP/PS}$ at $V = 3.3$ V

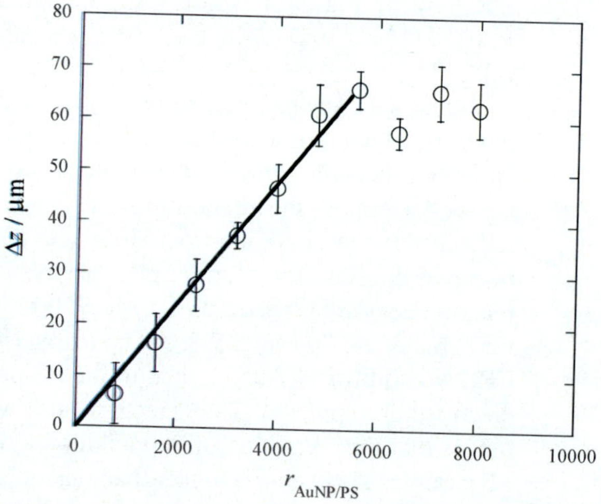

4.5 Quantification of Dissolved Biotin

The method discussed above is applicable to the determination of inhibitors for the avidin–biotin reaction. Biotin, selected as the model target in this system, is an important vitamin involved in significant metabolic reactions and biochemical processes. Biochemical disorders in animals, such as reduced carboxylase activity,

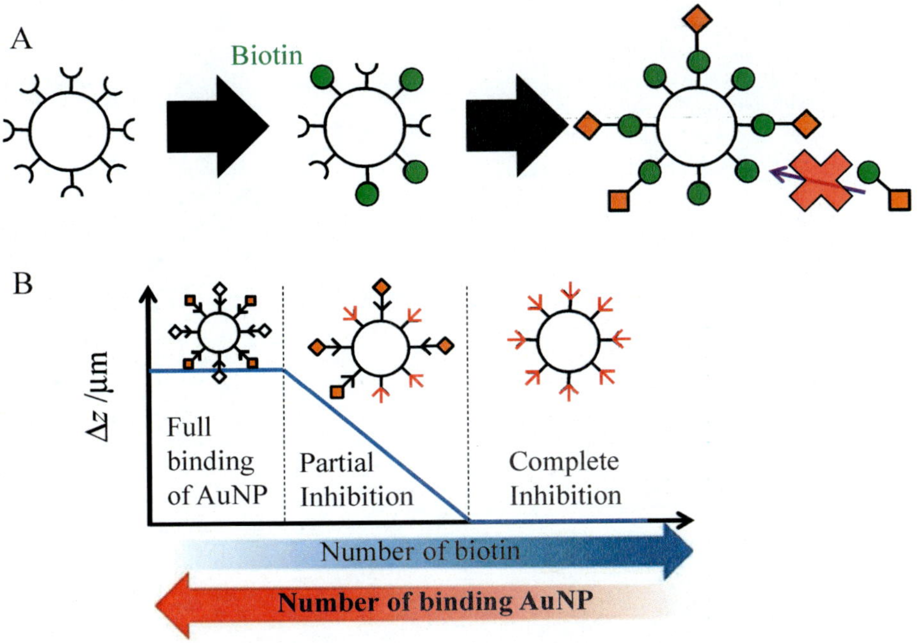

Fig. 4.14 Schematic of biotin quantification

inhibition of protein and RNA synthesis, and reduced antibody production, are caused by biotin deficiency [6]. Therefore, the sensitive determination method for biotin is crucial because it may affect human health. Many methods relying on liquid chromatography [7–10] or enzymatic reactions [11–13] have been developed.

The principle for biotin quantification based on the present method is represented in Fig. 4.14. The avidin-modified PS particles were treated with a biotin solution. Biotin molecules occupy the binding sites of avidin molecules anchored on the PS microparticles. Then, biotin-modified AuNPs react with the resultant PS particles. When the number of biotin molecules bound on a PS particle ($r_{B/PS}$) while it is small, AuNPs can bind to the PS particles until the maximum surface coverage is reached. However, as $r_{B/PS}$ increases, the biotin molecules block more binding sites and the number of unoccupied sites decreases. This reduces the binding of biotin-modified AuNPs and leads to a decrease in Δz. When biotin molecules in the solution completely occupy all the binding sites, biotin-modified AuNPs no longer bind to PS, and Δz becomes zero. Thus, biotin can be determined by this competitive binding.

Figure 4.15 compares the relationships between Δz and $r_{B/PS}$ for reaction times of 2 and 4 h. When $r_{B/PS}$ increases, Δz becomes smaller because the biotin molecules in the solution occupy the binding sites of avidin molecules and inhibit the subsequent binding of AuNPs. When $r_{B/PS} = 4.3 \times 10^6$, AuNP binding is completely inhibited, and Δz becomes zero regardless of the reaction time. The number of avidin molecules anchored on a PS particle was determined to be 1.09×10^6 using biotin-4-fluorescein. Because an avidin molecule has four sites for biotin binding, the number of binding

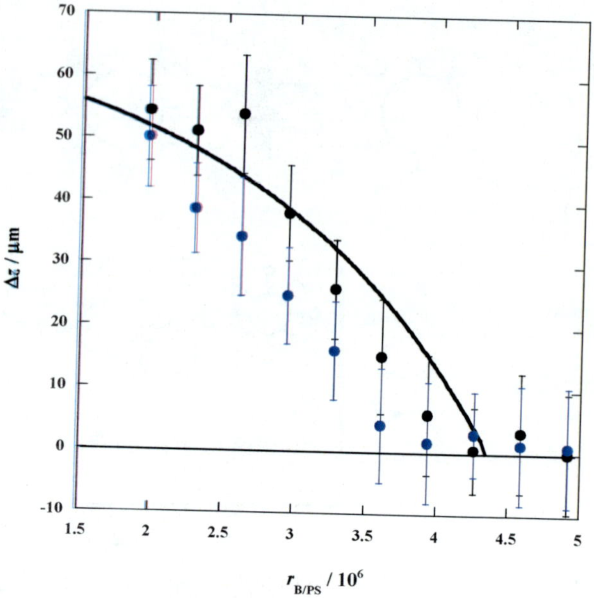

Fig. 4.15 Relationship between Δz and $r_{B/PS}$ for biotin anchored on AuNP with avidin anchored on PS and different reaction times; 2 h (blue) and 4 h (black)

sites per PS particle is 4.36×10^6, which agrees with $r_{B/PS} = 4.3 \times 10^6$ for the complete inhibition of AuNP binding.

The relationships between Δz and $r_{B/PS}$ were interpreted based on reaction kinetics. The black curve in Fig. 4.15 represents the result of curve-fitting using Eqs. (4.5)–(4.9) substituting $k_1 = 4.0 \times 10^5$ M^{-1} s^{-1} and $k_{-1} = 1.0 \times 10^{-4}$ s^{-1} as fitting parameters. The experimental values are adequately explained by the curve calculated by this kinetic model. Linear relationships are observed between $r_{B/PS}$ and Δz in the ranges of $r_{B/PS} = 2.0$–4.0×10^6 and 2.6–4.0×10^6 for the reaction times of 2 and 4 h, respectively. This indicates that the detection range can be adjusted by changing the reaction time. In this scheme, the detection limit was 3.6×10^6 biotin molecules in the solution.

The relationship between Δz and the biotin concentration (c_B) at $r_{B/PS} = 4.3 \times 10^6$ and $r_{AuNP/PS} = 4000$ was investigated to verify the reaction kinetics. As shown in Fig. 4.16, [14] at $c_B = 50$–500 pM, Δz approaches zero, and AuNP-binding is completely inhibited by binding biotin molecules. In contrast, Δz becomes larger for lower c_B, indicating that the biotin inhibition for AuNP binding is incomplete. The reaction of biotin molecules with avidin molecules anchored on PS particles did not reach equilibrium within a reaction time of 4 h at $c_B = 50$ pM or lower. Therefore, the concentration-based detection limit under this condition is 50 pM. The rate constants for the reaction of biotin molecules with the avidin-modified PS particles were calculated by curve-fitting to be $k_1 = 2.0 \times 10^8$ M^{-1} s^{-1} and $k_{-1} = 1.0 \times 10^{-4}$ s^{-1}. The association constant ($K = k_1/k_{-1}$) was calculated to be 2.0×10^{12} M^{-1}. This value is three orders of magnitude smaller than that of the molecular-based avidin–biotin reaction in the solution. This difference is attributed to the change in the diffusivity due to the modification of avidin and biotin to the PS

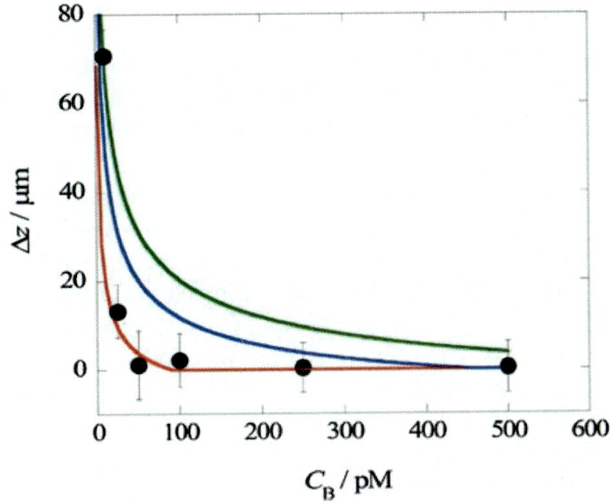

Fig. 4.16 Relationship between Δz and the biotin concentration (c_B) at a constant $r_{B/PS} = 4.3 \times 10^6$ and $r_{AuNP/PS} = 4000$. The reaction time was 4 h. Rate constants for the reaction of the avidin-modified PS microparticle with biotin were assumed; green, $k_1 = 2.00 \times 10^7$ M^{-1} s^{-1} and $k_{-1} = 1.00 \times 10^{-4}$ s^{-1}; blue, $k_1 = 4.00 \times 10^5$ M^{-1} s^{-1} and $k_{-1} = 1.00 \times 10^{-4}$ s^{-1}; red, $k_1 = 2.00 \times 10^8$ M^{-1} s^{-1} and $k_{-1} = 1.00 \times 10^{-4}$ s^{-1}. Reproduced from Ref. 14. Copyright 2018 American Chemical Society

particle and AuNP, respectively. The diffusion coefficient of molecules anchored on the particle is smaller than that of the free molecules in the solution, as discussed above. The rate constant is proportional to the diffusion constant of the reactants. The size of a PS particle (10 µm) is four orders of magnitude larger than that of an avidin molecule. The diffusion coefficient is also smaller for the PS particle by the same order. Thus, the difference in K between the interparticle and intermolecular systems is reasonable.

4.6 Summary and Conclusion

The CAG field recognizes the acoustic properties of a particle (density and compressibility) and levitates particles with different acoustic properties at different z values. The density change of the microparticles was induced by AuNP binding. In this chapter, a detection scheme based on the levitation coordinate shift was designed for the avidin–biotin reaction, which mediates the interparticle binding.

The density effect on z was investigated using gold-plated PMMA microparticles. The levitation behavior of gold-plated PMMA with different layer thicknesses (l) was followed by the calculation using Eq. (2.70). The linear relationship between Δz and the increase in density was confirmed (Fig. 4.7). The detection limit of l was 6.3 nm.

A linear relationship was confirmed between Δz and $r_{AuNP/MS}$, which provided the basis of methodological designs for quantifying various targets. The reaction kinetics for the reaction of avidin-modified PS microparticles with biotin-modified AuNPs were analyzed using Eqs. (4.5)–(4.9). The rate constants were determined to be $k_1 = 4.0 \times 10^5$ M^{-1} s^{-1} and $k_{-1} = 1.0 \times 10^{-4}$ s^{-1}. Thus, the association constant in this interparticle reaction was calculated to be $K = 4.0 \times 10^9$ M^{-1}, which was lower than that of the solution-phase association ($K = 10^{15}$ M^{-1}). This is caused by anchoring of the avidin and biotin molecules on the particles.

The proposed method was applied to biotin quantification in solution. Biotin molecules act as inhibitors for the interparticle reaction. With the increase in c_B, Δz decreases because the binding of biotin-anchored AuNPs is hindered by dissolved biotin molecules, which occupy the reaction sites of avidin molecules anchored on the PS microparticles. The detection range can be controlled by changing the reaction time. The determination limit was 50 pM. This result indicates that other reaction inhibitors can also be detected using this system.

References

1. Charcosset C, Limayenm I, Fessi H (2004) The membrane emulsification process—a review. J Chem Technol Biotechnol 79:209–218
2. Green NM, Joyson MA (1970) A preliminary crystallographic investigation of avidin. Biochem J 118:71–72
3. Waner MJ, Mascotti DP (2008) A simple spectrophotometric streptavidin-biotin binding assay utilizing biotin-4-fuluorescein. J Biochem Biophys Methods 70:873–877
4. Kada G, Falk H, Gruber HJ (1999) Rapid estimation of avidin and streptavidin by fluorescence quenching or fluorescence polarization. Biochim Biophys Acta 1427:44–48
5. Haiss W, Thanh NTK, Aveyard J, Fernig DG (2007) Anal Chem 79:4215–4221
6. Livaniou E, Costopoulou D, Vassiliadou I, Leondiadis L, Nyalala JO, Ithakissios DS, Evangelatos GP (2000) Analytical techniques for determining biotin. J Chromatogr A 881:331–343
7. Lahely S, Ndaw S, Arella F, Hasselamann S (1999) Determination of biotin in foods by high-performance liquid chromatography with post-column derivatization and fluorimetric detection. Food Chem 65:253–258
8. Nojiri S, Kamata K, Nishijima M (1998) Fluorescence detection of biotin using post-culumn derivatization with OPA in high performance liquid chromatography. J Pharm Biomed Anal 16:1357–1362
9. Staggs CG, Sealey WM, McCabe BJ, Teague AM, Mock DM (2004) Determination of biotin content of select foods using accurate and sensitive HPLC/avidin binding. J Food Compo Anal 17:767–776
10. Yomota C, Ohnishi Y (2007) Determination of biotin following derivazation with 2-nitrophenylhydrazine by high-performance liquid chromatography with on-line UV detection and electrospray-ionization mass spectrometry. J Chromatogr A 1142:231–235
11. Chang YS, Wu CH, Chang RJ, Shiuan D (1994) Determination of biotin concentration by a competitive enzyme-linked immunosorbent assay (ELISA) method. J Biochem Biophys Methods 29:321–329
12. Lu B, Iwuoha EI, Smyth MR, O'Kennedy R (1997) Development of an "electrically wired" amperometric immunosensor for the determination of biotin based on a non-diffusional redox osmium polymer film containing an antibody to the enzyme label horseradish peroxidase. Anal Chim Acta 345:59–66

13. Mishra S, Storer MK, Sherwin CMT, Lewis JG (2005) A simple binding assay for the direct determination of biotin in urine. Clin Chim Acta 360:60–66
14. Miyagawa A, Harada M, Okada T (2018) Zeptomole detection scheme based on levitation coordinate measurements of a single microparticle in a coupled acoustic-gravitational field. Anal Chem 90:2310–2316

Chapter 5
Label-Free Detection for DNA/RNA Molecules

Abstract In Chapter 4, a zmol-sensing scheme based on the levitation behavior of a single microparticle is proposed. However, this detection scheme needs the anchoring of target molecules on the microparticle or AuNP, which may lead to a change in the chemical and/or physical properties of the target molecules. In this chapter, a label-free detection scheme using DNA/RNA sandwich hybridization is demonstrated. In Sect. 5.2, the effect of the bp number, involved in two direct complementary hybridizations, on sensitivity is investigated before the sandwich hybridization scheme is examined. Based on the knowledge obtained in Sect. 5.2, detection schemes are designed to quantify the HIV-2 DNA and single nucleotide polymorphism by sandwich hybridization in Sect. 5.3. This approach is applied to the quantification of multiple microRNA (miRNA) molecules in Sect. 5.4.

Keywords Label-free · HIV-2 · MiRNA · Zmol detection

5.1 Experimental

Probe DNA molecules, miR-21 and miR-122, were purchased from Fasmac Co., Ltd. (Kanagawa, Japan). Total RNA extracted from human liver tumor tissue (lot no. A 505,425) was purchased from Cosmo Bio Co., Ltd. (Tokyo, Japan). Pre-miR-21 was purchased from Integrated DNA Technology (Illinois, USA). The sequences of miR-21-5p, miR-122-5p, and the premiR-21 were obtained from miRbase Website [1]. Tables 5.1, 5.2, 5.3 and 5.4 summarize the sequence of nucleotides used for DNA and RNA detections. The carboxyl-functionalized PMMAs with 6.33 ± 0.12 and 9.57 ± 0.21 μm were purchased from Microparticle GmbH (Berlin, Germany). The concentrations of these particles in stock solutions were determined to be 1.88×10^8 and 1.32×10^9 mL^{-1}, respectively, using the same method as described in Sect. 4.1.

Amino-terminated capture and reporter DNA were covalently conjugated onto the carboxyl-functionalized PMMA microparticles and AuNPs by the condensation reaction using 1-[3-(dimethylamino)propyl]-3-ethyl carbodiimide (EDC) and N-hydrozysuccinimide (NHS), respectively. The concentrations of EDC and NHS in MES buffer were 30 and 36 mg mL^{-1}, respectively. PMMA particle suspension of 100

© The Author(s), under exclusive license to Springer Nature Singapore Pte Ltd. 2021

A. Miyagawa, *Acoustic Levitation-Based Trace-Level Biosensing*, Springer Theses,
https://doi.org/10.1007/978-981-16-1425-5_5

Table 5.1 Sequences of nucleotides used for the direct binding of AuNPs with PMMA through complementary DNA hybridization

DNA	Sequence (5′ → 3′)	Functional group
Capture DNA	GCAACTAAATTCA-PMMA	Amino (3′)
12-nt	GAATTTAGTTGC-AuNP	Amino (3′)
11-nt	AATTTAGTTGC-AuNP	Amino (3′)
10-nt	ATTTAGTTGC-AuNP	Amino (3′)
9-nt	TTTAGTTGC-AuNP	Amino (3′)
8-nt	TTAGTTGC-AuNP	Amino (3′)
7-nt	TAGTTGC-AuNP	Amino (3′)

Table 5.2 Sequences of nucleotides used for detecting HIV-2 by sandwich hybridization

DNA	Sequence (5′ → 3′)	Functional group
Capture DNA	GCAACTAAATTCA-PMMA	Amino (3′)
Reporter DNA	AuNP-AAAGGACCAGGC	Amino (5′)
DNAs from HIV-2		
25-nt	TGAATTTAGTTGCGCCTGGTCCTTT	–
22-nt	AATTTAGTTGCGCCTGGTCCTT	–
20-nt	ATTTAGTTGCGCCTGGTCCT	–

Table 5.3 Sequences of nucleotides used for detecting single nucleotide polymorphisms (SNPs)

DNA	Sequence (5′ → 3′)	Functional group
Capture DNA	TGGCGTAGGCAA-PMMA	Amino (3′)
Reporter DNA	AuNP-TGGAGCTGG	Amino (5′)
Target DNA		
Wild DNA	TTGCCTACGCCACCAGCTCCAACT	
Mutant DNA	TTGCCTACGCCATCAGCTCCAACT	

μL (1.88×10^7 mL^{-1}) was added to 200 μL EDC/NHS solution to avoid hydrolysis immediately after buffer preparation. The resultant solution was incubated for 30 min at room temperature. This solution was added to 1 mL PBST (phosphate-buffered saline +0.05% Tween 20) solution and vortexed. Unreacted NHS and byproducts were removed by centrifugation at 6600 rpm. The amino-terminated capture DNA was added to the PMMA suspension and incubated for 4 h at room temperature with mixing. The amount of DNA on the surface of a PMMA particle was determined to be 1.6×10^7 mol using fluorescein-terminated complementary DNAs. Various probe DNAs were similarly anchored onto AuNPs.

For the direct binding method (Fig. 5.1a), DNA-anchored PMMA microparticles

Table 5.4 Sequences of nucleotides used for detecting miRNA

Symbol	Sequence (5′ → 3′)	Functional group
DNA		
miR-21 capture probe DNA	PMMA-TCAACATCAGT	Amino (5′)
miR-122 capture probe DNA	PMMA-CAAACACCATT	Amino (5′)
miR-21 reporter probe DNA	CTGATAAGCTA-AuNP	Amino (3′)
RNA		
miR-21	UAGCUUAUCAGACUGAUGU UGA	–
miR-122	UGGAGUGUGACAAUGGUG UUUG	–
Pre-miR-21	UGUCGGGUAGCUUAUCAGAC UGAUGUUGACUGUUGAAUCU CAUGGCAACACCAGUCGAUG GGCUGUCUGACA	–

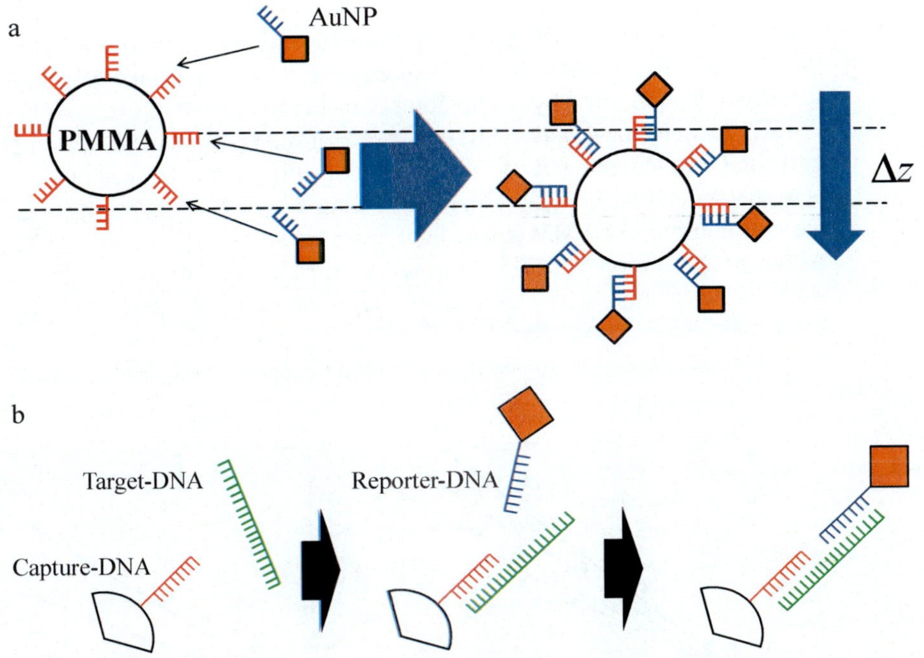

Fig. 5.1 Schematic representation of **a** direct binding between DNA-modified PMMA and AuNP and **b** sandwich hybridization for label-free detection

reacted with complementary DNA-anchored AuNPs. AuNP-bound PMMA particles were prepared by direct interparticle hybridization.

For the sandwich hybridization method (Fig. 5.1b), AuNP-bound PMMA particles were prepared by mixing the target DNA/RNA, probe DNA-anchored PMMA microparticles, and probe DNA-anchored AuNPs. The target DNA or RNA molecules were added to the DNA-anchored PMMA particles and incubated for 4 h at room temperature with mixing. The resultant suspension was added to the DNA-anchored AuNPs and incubated for 4 h.

The PMMA and AuNP-bound PMMA particles were observed using field-emission scanning electron microscopy (FE-SEM, JEOL, Japan). High-resolution images were obtained using secondary electron detection at an operating voltage of 6 kV. Samples were coated with Pt/Pd.

5.2 Effect of the Base Pair Number on Sensitivity

Figure 5.1a shows the schematic representation of the formation of AuNP-bound PMMA particles through the direct interparticle DNA hybridization. In this study, sensitivity is influenced by the bp number involved in the hybridization between the PMMA particle and AuNP. Thus, the effect of the bp number on sensitivity can be evaluated by measuring Δz in the CAG field. Figure 5.2a shows the relationships between Δz and $r_{AuNP/PMMA}$ obtained with the direct DNA hybridization, in which the concentration of PMMA was kept constant at $n_{PMMA} = 1.4 \times 10^4$ mL^{-1} [2]. Although Δz is proportional to $r_{AuNP/PMMA}$ for all bp numbers except for 7 bp, the slope of the Δz-$r_{AuNP/PMMA}$ plots decreases with decreasing bp number because the DNA hybridization becomes weak. For 7 bp, the change in Δz with $r_{AuNP/PMMA}$ was not detected, suggesting that no hybridization occurred or the duplex was dissociated

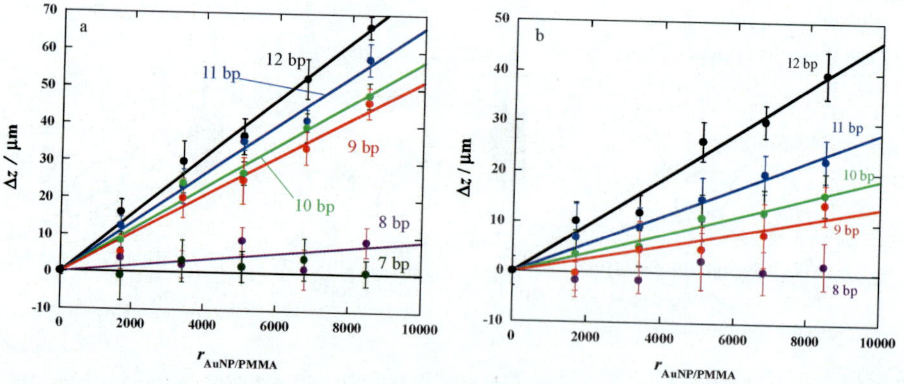

Fig. 5.2 Relationships between Δz of AuNP-bound PMMA and $r_{AuNP/PMMA}$ in direct interparticle DNA hybridization. **a** $n_{PMMA} = 1.4 \times 10^4$ and **b** 65 mL^{-1}. Reproduced from Ref. 2. Copyright 2018 American Chemical Society

before measurements. These results indicate that the sensitivity in this system can be controlled by changing the bp number. The detection limit was 400, 800, 900, 2000 and 5200 $r_{AuNP/PMMA}$ for 12, 11, 10, 9, and 8 bp, respectively. The binding affinity in the interparticle DNA hybridization also depends on the particle concentration. Compared to the results for $n_{PMMA} = 1.4 \times 10^4$ mL^{-1}, Δz decreased for $n_{PMMA} =$ 65 mL^{-1} as shown in Fig. 5.2b [2]. For 8 bp, the slope of the Δz-$r_{AuNP/PMMA}$ plot becomes zero in the latter case. Thus, the sensitivity in this detection scheme can also be controlled by adjusting n_{PMMA}.

Table 5.5 summarizes the slopes of the Δz—$r_{AuNP/PMMA}$ plots for 8–12 bp at $n_{PMMA} = 1.4 \times 10^4$ and 65 mL^{-1}. The ratios of the slope for 12 bp to those for the other bp (s_{12}/s_{bp}) are also listed in this table. The s_{12}/s_{bp} ratios for 9–12 bp were not very different at $n_{PMMA} = 1.4 \times 10^4$ mL^{-1}, indicating that these DNA lengths are not clearly distinguishable under this condition. In contrast, the difference in the s_{12}/s_{bp} between 8 and 9 bp was evidently larger than other ratios. In addition, the selectivity of this method to the DNA length could be increased by decreasing n_{PMMA}. The s_{12}/s_9 and s_{12}/s_{10} values at $n_{PMMA} = 65$ mL^{-1} were twice as large as those at $n_{PMMA} = 1.4 \times 10^4$ mL^{-1}. Thus, selectivity can be enhanced by decreasing n_{PMMA} and adjusted by changing the n_{PMMA}.

The dissociation constant for dsDNA, K_{DNA}, for each bp can be calculated from $\Delta G°$, which is given by the NN model proposed by Walder and co-workers [3]. Figure 5.3 shows the relationships between calculated K_{DNA} and Δz measured at $r_{AuNP/PMMA} = 8.50 \times 10^3$ [2]. Black and red symbols show the results obtained at $n_{PMMA} = 1.4 \times 10^4$ and 65 mL^{-1}, respectively. Δz decreases as the bp number decreases in a sigmoidal-shaped curve because of the increase in K_{DNA}. In addition, Δz decreases with the decrease in n_{PMMA} for all cases because the probe DNA concentration, which is proportional to n_{PMMA}, decreases. The solid curves in Fig. 5.3 show the change in Δz for each bp predicted from the K_{DNA} and DNA concentrations.

Table 5.5 Slopes of the Δz-$r_{AuNP/PMMA}$ plot for 8–12 bp and ratios of slope for 12 bp to slopes for other bps

n_{PMMA} / mL^{-1}	bp	Slope (s)/10^{-3} µm	$\sigma^a/10^{-3}$ µm	s_{12}/s_{bp}	$\sigma/10^{-3}$ µm
1.4×10^4	12	7.8	0.57	–	–
	11	6.6	0.50	1.18	0.85
	10	5.7	0.57	1.37	0.73
	9	5.2	0.55	1.50	0.67
	8	1.2	0.85	6.50	0.15
65	12	4.6	0.58	–	–
	11	2.8	0.29	1.64	0.61
	10	1.9	0.22	2.42	0.41
	9	1.3	0.47	3.54	0.28
	8	0.1	0.35	46.0	0.02

a, standard deviation ($n = 5$)

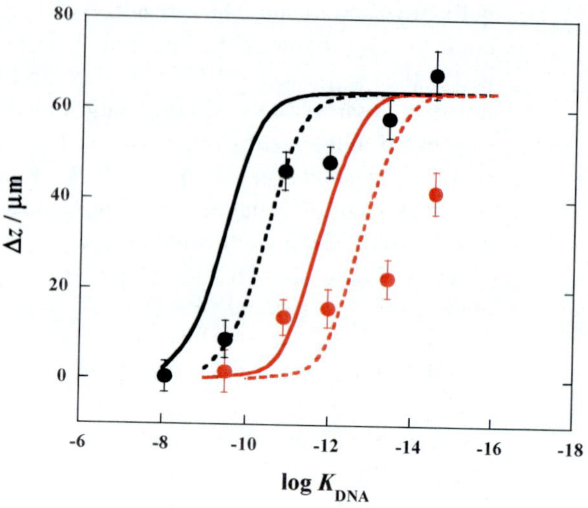

Fig. 5.3 Relationships between the calculated K_{DNA} and Δz for each bp at $r_{AuNP/PMMA} = 8.50 \times 10^3$. Black curves $n_{PMMA} = 1.4 \times 10^4$ mL^{-1} and red curves $n_{PMMA} = 65$ mL^{-1}. Solid curves were determined using K_{DNA} calculated with the NN model proposed by Walder and co-workers [3]. Dashed curves were calculated assuming K_{DNA} smaller by 1.5 orders of magnitude than the above values. Reproduced from Ref. 2. Copyright 2018 American Chemical Society

It was assumed that Δz was proportional to $r_{AuNP/PMMA}$, and the maximum Δz was 65 μm. The solid curves predict the sigmoidal changes of Δz with $r_{AuNP/PMMA}$ for both $n_{PMMA} = 1.4 \times 10^4$ and 65 mL^{-1}, but do not exhibit good agreement with the experimental values.

The anchoring of DNA on a particle may affect the thermodynamics of DNA hybridization. Fong et al. reported that hybridization is promoted when DNAs are anchored on the nanoparticle surface because of the enthalpic preference of the DNA structure for subsequent hybridization [4]. In contrast, the rate constant of a diffusion-limited reaction depends on the diffusion coefficients of the reactants. The diffusion coefficient is reduced by anchoring molecules on a particle, as discussed in Sect. 4. The reaction rate may decrease by two orders of magnitude because the diameter of AuNPs is larger than the size of DNA molecules by two orders of magnitude. The binding constant should also be lowered by this effect. Thus, two opposite effects are involved in interparticle DNA hybridization between PMMA microparticles and AuNPs. The difference between the experimental and calculated Δz values is caused by these effects. The dashed curves in Fig. 5.3 were calculated assuming K_{DNA} values smaller by 1.5 orders of magnitude than the corresponding values calculated from ΔG°. The agreement between the experimental and calculated Δz values is considerably better for this assumption. Thus, the overall K_{DNA} becomes smaller by 1.5 orders of magnitude by DNA anchoring on the particles.

5.3 Label-Free DNA Sensing by Sandwich Hybridization

5.3.1 *Detection of HIV-2 DNA*

Label-free DNA detection was performed using sandwich hybridization. The specific base sequence of HIV-2 was chosen as the model target DNA (its sequence is shown in Table 5.1). Figure 5.1b shows a schematic representation of the detection scheme using sandwich hybridization. Capture-DNA (13 nt) and reporter DNA (12 nt) were anchored on the PMMA microparticles and AuNPs, respectively. The DNA undergoes complementary hybridization with the target DNA to form the 25 bp DNA duplex. When the bp number exceeds 12 (log K_{DNA} ~ 14.5), the dsDNA formed by complementary hybridization can be regarded as indissociable during the present detection scheme, as discussed in Sect. 5.2. DNA-anchored PMMAs were first treated with target DNA ($r_{DNA/PMMA}$ = 3000–15,000). Then, the resultant particles were treated with DNA-anchored AuNPs at a constant $r_{AuNP/PMMA}$ of 1.55×10^4. Figure 5.4a shows the relationship between Δz and $r_{DNA/PMMA}$ at $n_{PMMA} = 9.4 \times 10^3$ mL^{-1} [2]. As seen in the figure, Δz increased linearly with increasing $r_{DNA/PMMA}$. At $n_{PMMA} = 9.4 \times 10^3$ mL^{-1}, 2370 HIV-2 DNA molecules were detectable, which corresponded to zmol label-free detection.

In Sect. 5.2, the effect of the bp number on Δz was investigated, demonstrating that the stability of DNA hybridization becomes weak as the bp number decreases. The effect of the bp number on the stability of the sandwich hybridization was also evaluated. In the above case, the 25 bp dsDNA was formed as a result of the sandwich hybridization to provide a stable binding between PMMA and AuNPs. The total bp number was reduced to 20 (10 + 10) and 22 bps (11 + 11), and Δz of the resulting AuNP-bound PMMA was measured. The results are summarized in Fig. 5.4a. The slope of Δz—$r_{DNA/PMMA}$ plots decreased with a decrease in the bp number, suggesting

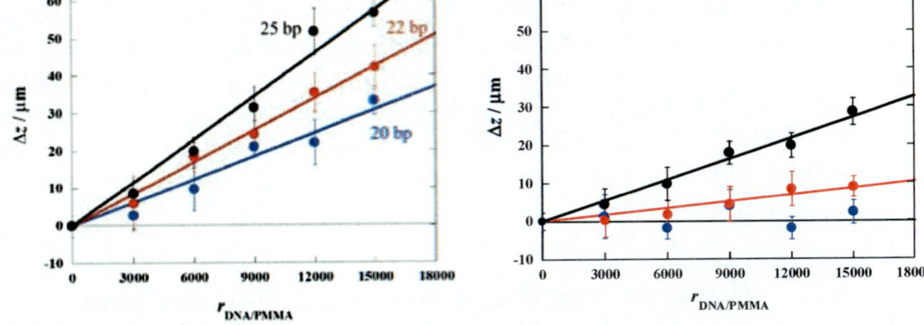

Fig. 5.4 Relationships between Δz of AuNP-bound PMMA and $r_{AuNP/PMMA}$ at n_{PMMA} = **a** 9.4 $\times 10^3$ and **b** 65 mL^{-1} in sandwich interparticle hybridization. Reproduced from Ref. 2. Copyright 2018 American Chemical Society

Table 5.6 Slopes of the Δz–$r_{DNA/PMMA}$ plot for 20–25 bps and ratios of the slope for 25 bps to those for other bps at $n_{PMMA} = 9.4 \times 10^3$ and 65 mL^{-1}

n_{PMMA}/mL^{-1}	bp	Slope (s)/10^{-3} µm	σ/10^{-3} µm	s_{bp}/s_{25}	σ/10^{-3} µm
9.4×10^3	25	3.8	0.35	–	–
	22	2.8	0.16	0.74	0.11
	20	2.1	0.28	0.55	0.16
65	25	1.8	0.15	–	–
	22	0.6	0.13	0.33	0.23
	20	0.0	0.23	–	–

that the DNA duplex was destabilized by a decrease in the number of bp involved in the hybridization. Figure 5.4b shows the effect of the PMMA concentration on the sensitivity, in which $n_{PMMA} = 65\ mL^{-1}$ [2]. The slope of Δz—$r_{DNA/PMMA}$ plot for 25 bp (s_{25}) at $n_{PMMA} = 65\ mL^{-1}$ was half of that at $n_{PMMA} = 9.4 \times 10^3$, and s_{22} was one-fourth of the corresponding value at $n_{PMMA} = 9.4 \times 10^3$. In the former condition, no Δz change was observed for the 20 bp duplex formation. Thus, the difference in Δz between n_{PMMA} values in sandwich hybridization was larger than that for direct binding. As two hybridizations are involved in sandwich hybridization, the concentration largely influenced the sensitivity compared to the direct binding system.

Table 5.6 summarizes the slopes of Δz-$r_{DNA/PMMA}$ plots for 20, 22, and 25 bp with $n_{PMMA} = 9.4 \times 10^3$ and 65 mL^{-1}. The ratios of the slope for 20 or 22 bps to those for 25 bps (s_{bp}/s_{25}) are also listed. The s_{bp}/s_{25} values at $n_{PMMA} = 9.4 \times 10^3\ mL^{-1}$ were smaller than those at $n_{PMMA} = 65\ mL^{-1}$; $s_{22}/s_{25} = 0.71$ and 0.33, and $s_{20}/s_{25} = 0.56$ and 0.0 at $n_{PMMA} = 9.4 \times 10^3$ and 65 mL^{-1}, respectively. This indicates that the selectivity is enhanced at lower n_{PMMA} values. Sandwich hybridization involves two separate hybridizations, that is, capture-target and reporter-target DNA hybridization. For example, 25 bps dsDNA formation involves 12- and 13-bps hybridizations. The selectivity in the sandwich hybridization should be discussed based on the combination of two hybridizations involved in the direct interparticle hybridization discussed in Sect. 5.2. Using the values in Table 5.5, s_{22}/s_{25} was calculated to be $(s_{11}/s_{12})(s_{11}/s_{12}) = 0.74$ at $n_{PMMA} = 9.4 \times 10^3\ mL^{-1}$ and 0.37 at $n_{PMMA} = 65\ mL^{-1}$. Similarly, s_{20}/s_{25} was calculated to be $(s_{10}/s_{12})(s_{10}/s_{12}) = 0.55$ at $n_{PMMA} = 9.4 \times 10^3\ mL^{-1}$ and 0.02 at $n_{PMMA} = 65\ mL^{-1}$. Thus, the selectivity in sandwich hybridization can be predicted from that in direct interparticle hybridization.

5.3.2 *Detection for a Single Nucleotide Polymorphism*

As discussed above, the sensitivity of this detection system can be controlled by adjusting the bp numbers in complementary hybridization and n_{PMMA}. Single nucleotide polymorphisms (SNPs) were detected based on this sensitivity control scheme. A part of the KRAS gene sequence was used as the target SNP. The KRAS gene encodes a GTP-binding protein, which is related to cell proliferation and tumor progression. Inhibition of epidermal growth factor receptor (EGFR) with monoclonal antibody is utilized for cancer patients with a wild-type KRAS gene [5]. However, the EGFR inhibitor does not act on patients with mutant KRAS gene, that is, SNP [6]. Thus, the identification of SNPs for the KRAS gene is needed to effectively promote cancer treatment.

Figure 5.5 shows the schematic representation of the principle of SNP detection. The capture DNAs with 12 nt were anchored on the PMMA microparticles to equally bind both wild and mutant DNA with sufficient stability by forming 12-bp dsDNA. The reporter DNAs with 9 nt on the AuNP were used for the identification of SNPs. Thus, the 9-bp dsDNA was formed for wild DNA, whereas the 8-bp dsDNA was formed between the mutant DNA and AuNP. In Sect. 5.2, a large difference in the affinity between 8 and 9 bp was indicated; 9 bp hybridization was detected by a levitation shift of PMMA, whereas 8 bp hybridization was not detectable. Therefore, wild DNA is selectively detectable using this principle. In addition, the selectivity between bp numbers was enhanced by decreasing n_{PMMA}, indicating that the detection selectivity of wild DNA against mutant DNA can be enhanced by reducing n_{PMMA}. Figure 5.6 shows the relationships between Δz and $r_{\mathrm{DNA/PMMA}}$ for SNP at $n_{\mathrm{PMMA}} = 9.4 \times 10^3$ and $1.17 \times 10^2\ \mathrm{mL}^{-1}$ [2]. Both wild and mutant DNAs were detectable at $n_{\mathrm{PMMA}} = 9.4 \times 10^3\ \mathrm{mL}^{-1}$, which indicates that SNP cannot be identified. In contrast, at $n_{\mathrm{PMMA}} = 1.17 \times 10^2\ \mathrm{mL}^{-1}$, wild DNA was detectable, whereas mutant DNA showed a zero slope. Thus, selective detection was accomplished by adjusting the bp number and n_{PMMA}.

Figure 5.7 compares the effect of n_{PMMA} on the slopes of Δz-$r_{\mathrm{DNA/PMMA}}$ plots for wild and mutant DNAs. Although the slope decreased as n_{PMMA} decreased for both DNAs that for the mutant was smaller than that for the wild DNA. The slope for mutant DNA was almost zero at $2.3 \times 10^2\ \mathrm{mL}^{-1}$, indicating that mutant DNA is not detectable in this condition. In contrast, the wild DNA was still detectable; the detection limit was 2700 at $n_{\mathrm{PMMA}} = 2.3 \times 10^2\ \mathrm{mL}^{-1}$. The dependence of Δz on

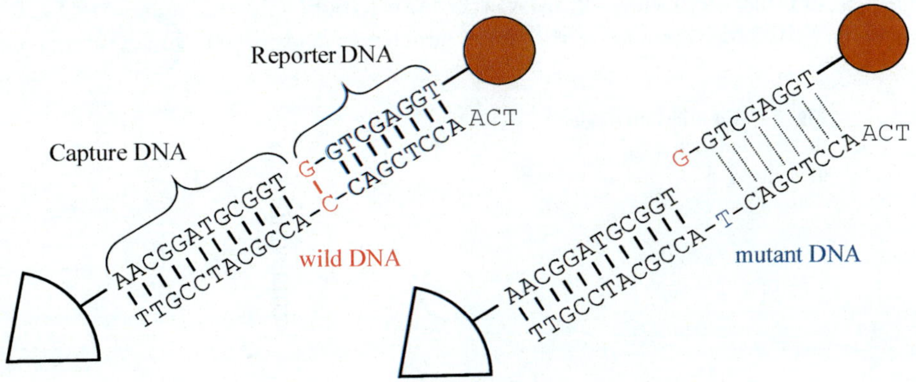

Fig. 5.5 Schematic representation of the SNP detection principle

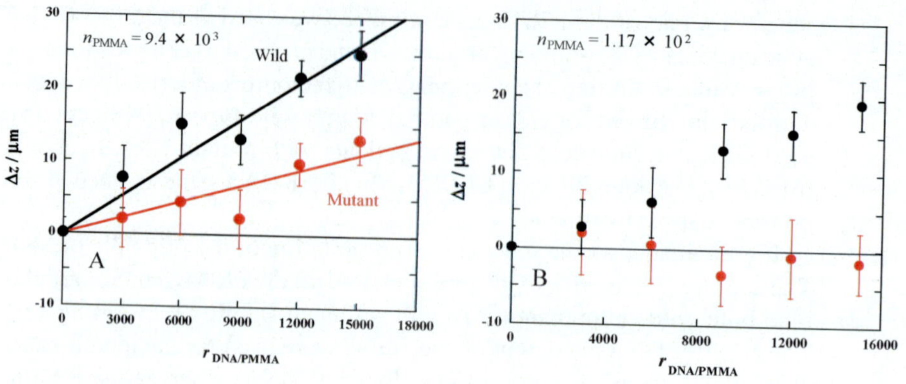

Fig. 5.6 Relationships between Δz of AuNP-bound PMMA and $r_{DNA/PMMA}$ for SNP at **a** $n_{PMMA} = 9.4 \times 10^3$ and **b** 1.17×10^2 mL^{-1} in the sandwich interparticle hybridization. Reproduced from Ref. 2. Copyright 2018 American Chemical Society

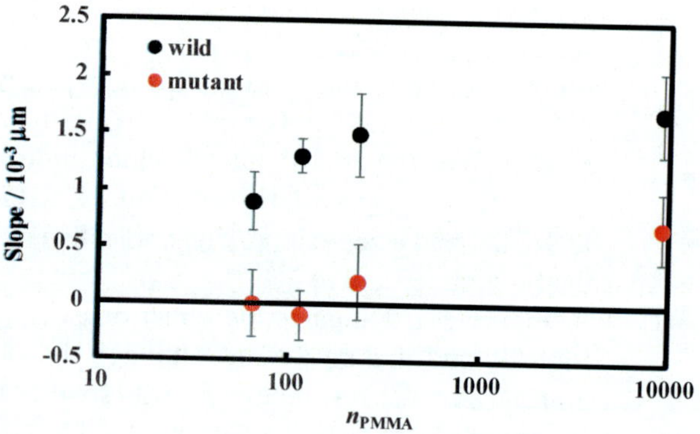

Fig. 5.7 Effect of n_{PMMA} on the slope of the Δz-$r_{DNA/PMMA}$ plot for wild and mutant DNAs

n_{PMMA} at $r_{DNA/PMMA} = 15{,}000$ for wild and mutant DNAs was analyzed based on the equilibrium model shown in Fig. 5.8. The association constant for the hybridization of probe DNA on the PMMA is common to both wild and mutant DNAs, K_1, = 3.47 × 10^{14} M^{-1}, = 3.47 × 10^{14} M^{-1}, which was determined from $\Delta G°$. In contrast, the

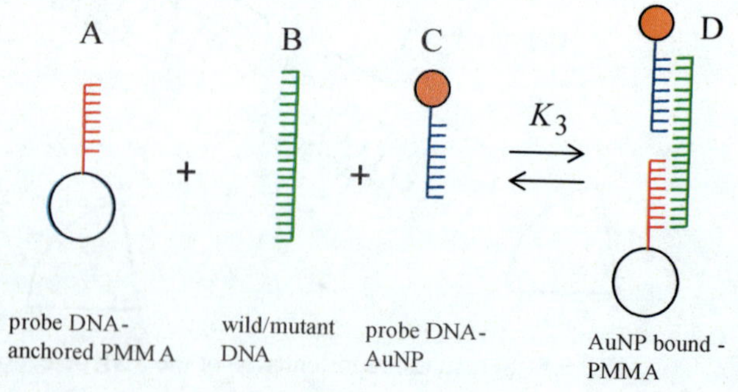

Fig. 5.8 Equilibrium model for the sandwich hybridization

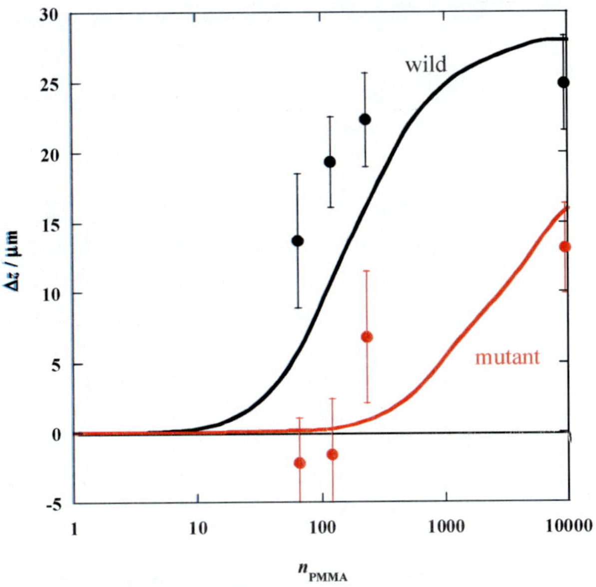

Fig. 5.9 Relationship between Δz and n_{PMMA} at $r_{DNA/PMMA} = 15{,}000$. Curves were calculated with $K_3 = 2.69 \times 10^{25}$ and 1.07×10^{24} M^{-2} for wild and mutant DNAs, respectively

association constants for the hybridization of probe DNA on AuNP, K_2, should be different for the wild and mutant DNAs, that is, $K_2 = 7.76 \times 10^{10}$ for the wild DNA and 3.00×10^9 M^{-1} for the mutant DNA. Thus, the equilibrium constants for the formation of AuNP-bound PMMA in Fig. 5.8 are given by $K_3 = K_1 \cdot K_2$, calculated to be 2.69×10^{25} and 1.07×10^{24} M^{-2} for wild and mutant DNAs, respectively. Figure 5.9 shows the relationship between Δz measured at $r_{DNA/PMMA} = 15{,}000$ and n_{PMMA}. Solid curves represent the results of the calculation using $K_3 = K_1 \cdot K_2 = 2.69 \times 10^{25}$ and 1.07×10^{24} M^{-2} for wild and mutant DNAs, respectively. The calculation was in good agreement with the experimental values. As discussed in Fig. 5.3, DNA anchoring on the particle made the association constant between the probe DNAs on the PMMA and AuNP larger than the corresponding value predicted from theoretical $\Delta G°$ by 1.5 orders of magnitude. However, in the sandwich hybridization system, the theoretical association constants explain the change in Δz, as shown in Fig. 5.9. DNA molecules were anchored on the particles in the direct hybridization, whereas the target DNA was free and not anchored on the particle in sandwich hybridization. This resulted in the difference in the association constant, which explains the change in Δz.

5.4 Multiple MicroRNA Quantification

5.4.1 *One-Pot Sample Preparation for Probing Multiple MicroRNA Molecules*

Figure 5.10 shows a schematic representation of one-pot sample preparation for multiple miRNA detection. DNA-anchored PMMA particles with different diameters (6.33 and 9.57 μm) were added to the sample solution, and then the mixture was shaken for 4 h. The sequence of probe DNA molecules on the 6.33 μm PMMA was selected to probe miR-122, whereas that on the 9.57 μm PMMA was designed to capture miR-21. AuNPs, on which reporter DNA molecules for miR-21 or miR-122 were anchored, were added to the resultant solution, and the mixture was shaken for 4 h. Because these PMMA particles can be identified on a microscopic view in terms of size, as shown in Fig. 5.11, [7] z for each microparticle was determined similarly as stated above.

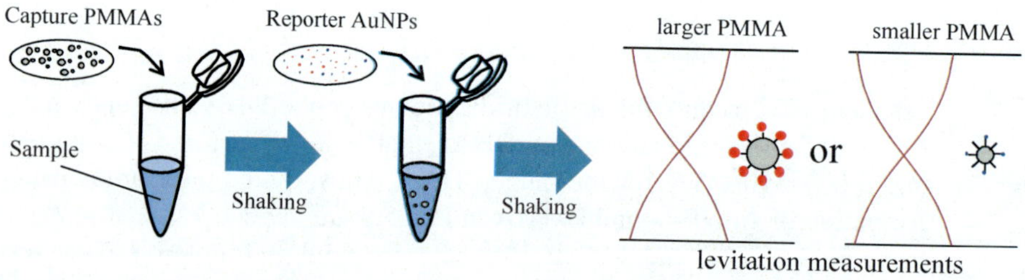

Fig. 5.10 Schematic representation of one-pot sample preparation for multiple miRNA detection

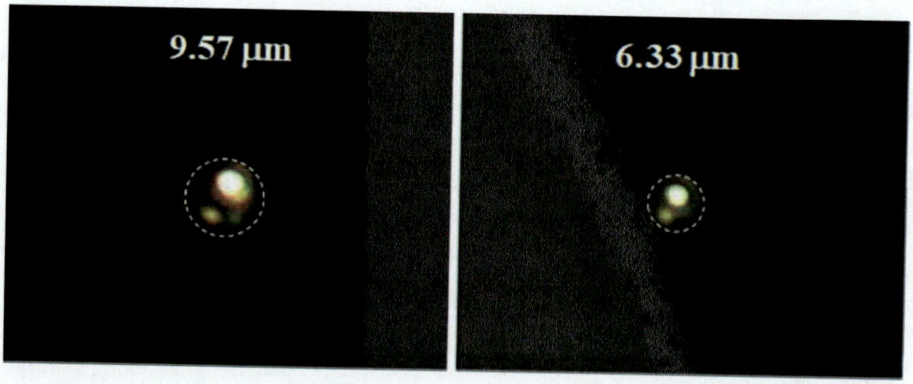

Fig. 5.11 Microscopic images of 9.57- (left) and 6.33 μm (right) PMMAs at the same magnification. Reproduced from Ref. 7. Copyright 2018 American Chemical Society

5.4.2 *Verification of Gold Nanoparticle Binding*

Figure 5.12 shows the images of an AuNP-bound 9.57 μm PMMA particle with varying $r_{RNA/PMMA}$ levitated in the CAG field [7]. In this experiment, miR-21 was selected as the target RNA. It is evident that z decreases as $r_{RNA/PMMA}$ increases. Thus, the sandwich detection scheme can also be applied to the miRNA.

The AuNP binding to PMMA microparticles by sandwich hybridization was verified by FE-SEM observation. Figure 5.13 shows FE-SEM images of 9.57 μm PMMA microparticles [7]. A number of AuNPs were observed on the PMMA surface after AuNP binding to PMMA microparticles through hybridization with miRNA. In contrast, only a few AuNPs were observed on the PMMA surface in the absence of miRNA (Fig. 5.13b) in the reaction system. AuNPs may adhere to PMMA during sample preparation. The coverage of AuNPs on the PMMA was calculated using backscattered electron images: 0.39 and 1.7% for $r_{RNA/PMMA}$ = 2000 and 10,000, respectively. Assuming that AuNPs completely bind to PMMA particles at $r_{RNA/PMMA}$ = 10,000, the coverage of AuNP should be 25%. Thus, the number of AuNPs on the PMMA in the SEM images was considerably smaller than the prediction from the complete binding of AuNPs. However, the coverage of AuNPs determined by SEM is correlated with the $r_{RNA/PMMA}$ value. Thus, AuNP binding to PMMA through sandwich hybridization with miRNA was confirmed by direct SEM observations, although the number of AuNPs was not consistent with the acoustic levitation measurements. The actual number of AuNP bound on PMMA microparticles is discussed in more detail in Sect. 6.2.

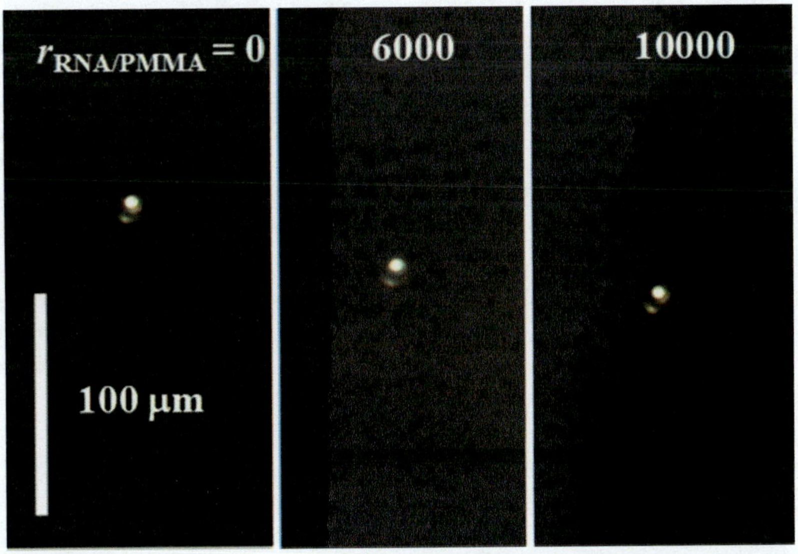

Fig. 5.12 Images of an AuNP-bound 9.57 μm PMMA particle levitated in the CAG field at varying $r_{RNA/PMMA}$. Reproduced from Ref. 7. Copyright 2018 American Chemical Society

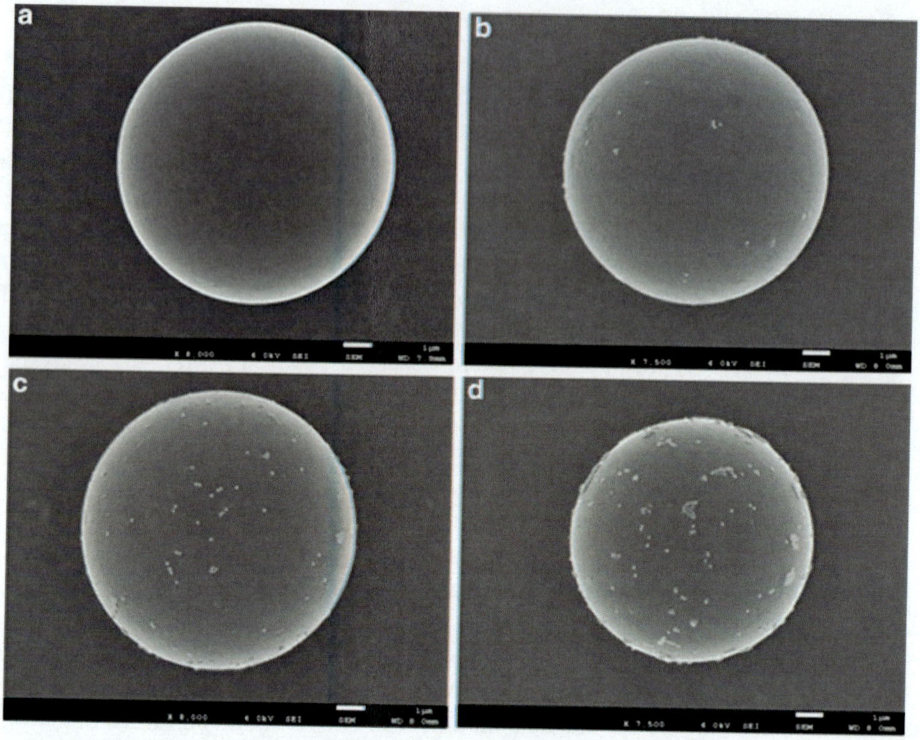

Fig. 5.13 FE-SEM images of 9.57 μm AuNP-bound PMMA microparticles. **a** Bare particle, **b** without miRNA, **c** with miRNA at $r_{RNA/PMMA} = 2000$, and **d** with miRNA at $r_{RNA/PMMA} = 10{,}000$. Reproduced from Ref. 7. Copyright 2018 American Chemical Society

In Fig. 5.13c and d, the islands of AuNPs are observed on the PMMA microparticles, and they were also observed on several other AuNP-bound PMMA microparticles. Ideally, AuNPs should be homogeneously distributed on the PMMA surface because AuNPs are terminated with amino groups and, therefore, the electrostatic repulsion between positively charged AuNPs should prevent their aggregation. The AuNP islands suggest that the carboxyl groups are not homogeneously distributed on the surface of the PMMA microparticles. The diameter of the AuNPs was 100 nm. When AuNPs bind to the area where carboxyl groups are densely present, nearby carboxyl sites cannot react with AuNPs. This possibly causes a decrease in the binding capacity.

5.4.3 *Quantification of MiR-21 and MiR-122*

Black plots and line in Fig. 5.14 show the relationship between Δz and $r_{RNA/PMMA}$ for miR-21 in the absence of interferents [7]. As can be seen, Δz is proportional to $r_{RNA/PMMA}$. The detection limit for miR-21 was 640 molecules. In general, biological

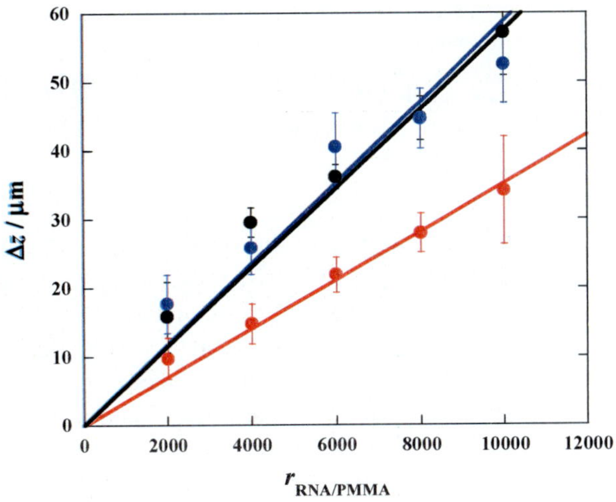

Fig. 5.14 Relationships between Δz and $r_{RNA/PMMA}$ for AuNP-bound PMMA prepared by the sandwich hybridization with miR-21. Black: miR-21 in standard solutions, red: miR-21 added in pre-miR-21 without annealing, and blue: miR-21 added in pre-miR-21 with annealing at 90 °C. Reproduced from Ref. 7. Copyright 2018 American Chemical Society

samples contain several types of off-target RNAs, such as SNPs and precursor RNAs. RNA polymerase II in the cell transcribes primary miRNAs, which are comprised of 1000–3000 bases. The RNase III enzyme Drosha and double-stranded RNA binding protein Pasha cleaves these miRNAs to form pre-miRNAs with stem-loop structures, which contain 70–100 nucleotides. The pre-miRNAs are further cleaved by RNase III enzyme Dicer into 18–24 oligonucleotides, which are mature miRNAs [8, 9]. Thus, pre-miRNAs, which have the same sequence as a mature miRNA, may act as an interferent, leading to overestimation or underestimation of the amount of mature miRNAs. When miR-21 was detected in the presence of tenfold excess of pre-miR-21, the slope of Δz–$r_{RNA/PMMA}$ plot was lowered (red plots and line in Fig. 5.14). This indicates that pre-miR-21 interacts with miR-21 and interferes with the hybridization between miR-21 and probe DNAs. The number of pre-miR-21 on the PMMA was estimated to be 4000 for $r_{RNA/PMMA} = 10{,}000$ from the difference in Δz between the black and red plots. When $r_{RNA/PMMA} = 10{,}000$, the equilibrium concentrations of probe DNA on the PMMA, pre-miR-21, and dsDNA between pre-miR-21 and probe DNA were calculated to be 5.1×10^{-8}, 3.0×10^{-10}, and 1.3×10^{-11} M, respectively. Thus, the association constant of pre-miR-21 with probe DNA on PMMA was determined to be 8.2×10^5 M^{-1}. As the 11 bps hybridization may occur between pre-miR-21 and probe DNA, an association constant of 1.8×10^{12} M^{-1} is expected from the NN model [3]. Thus, pre-miR-21 forms an incomplete duplex with the probe DNA on the PMMA because it adopts a more stable structure (Fig. 5.15) [7].

$\Delta G°$ for the stem-loop structure formation of pre-miR-21 (Fig. 5.15) was calculated to be −41.07 kcal mol^{-1} using the same method as described in Sect. 5.2. Similarly, the hybridizations between miR-21 and pre-miR-21, miR-21 and the capture DNA, and miR-21 and reporter DNA were calculated to be −30.91, −16.7, and −16.8 kcal mol^{-1}, respectively. Thus, the hybridization between pre-miR-21 and

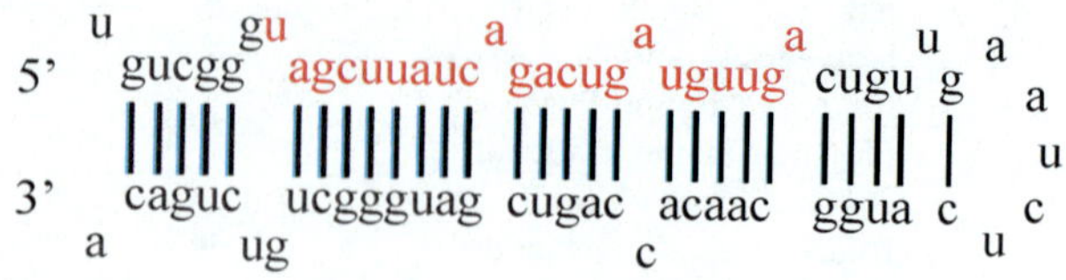

Fig. 5.15 Structure of pre-miR-21. The red letters represent the sequence of mature miR-21. Reproduced from Ref. 7. Copyright 2018 American Chemical Society

miR-21 is stronger than that between miR-21 and capture DNA or reporter DNA. Therefore, pre-miR-21 showed a negative interference.

After the capture DNA-anchored PMMAs were added to the solution containing pre-miR-21 and miR-21, the solution was annealed at 90 °C for 1 h to facilitate the formation of the proper stem-loop structure of pre-miR-21 and to prevent the interaction between pre-miR-21 and miR-21. After this procedure, reporter DNA-anchored AuNPs were added to the solution. The results are shown in Fig. 5.14 as blue plots and lines. The original Δz values obtained without pre-miR-21 were recovered by employing the annealing process. Pre-miR-21 forms a stable stem-loop structure and does not interact with miR-21 after annealing. Thus, the annealing step eliminates the interference from pre-miRNA.

The determination of multiple miRNAs in the total RNA extracted from liver cancer tissues was performed using PMMA microparticles with two different diameters. The sample preparation was performed using the one-pot method shown in Sect. 5.4.1 (see Fig. 5.10). Both miR-21 and miR-122 in total RNA (100 mg mL^{-1}) were quantified by the standard addition method. All samples were annealed at 90 °C for 1 h. Figure 5.16 shows the relationships between the spiked concentration of miR-21 or miR-122 (c_{RNA}) and Δz [7]. The miR-21 and miR-122 in the total RNA were determined to be 0.31 and 0.88 amol ng^{-1}, respectively. For miR-21, Chan et al. reported 0.43 amol ng^{-1} using fluorescence detection [10] and Feng et al. reported 0.48 amol ng^{-1} using the isothermal amplification technique [11]. These concentrations are in good agreement with the one obtained in the present study. No reported values were found for miR-122 to the best of the author's knowledge. Thus, the present scheme successfully worked for the simultaneous detection of multiple miRNA detection molecules.

5.5 Summary and Conclusion

A label-free detection scheme was demonstrated using sandwich hybridization. The target DNA/RNA was entrapped by probe DNAs anchored on the PMMA microparticles. The AuNP binding to the PMMA microparticles via hybridization with target DNA/RNA induced the density change of microparticles, which were detected in the CAG field. In this chapter, HIV-2, SNP, and miRNA were successfully detected using this approach.

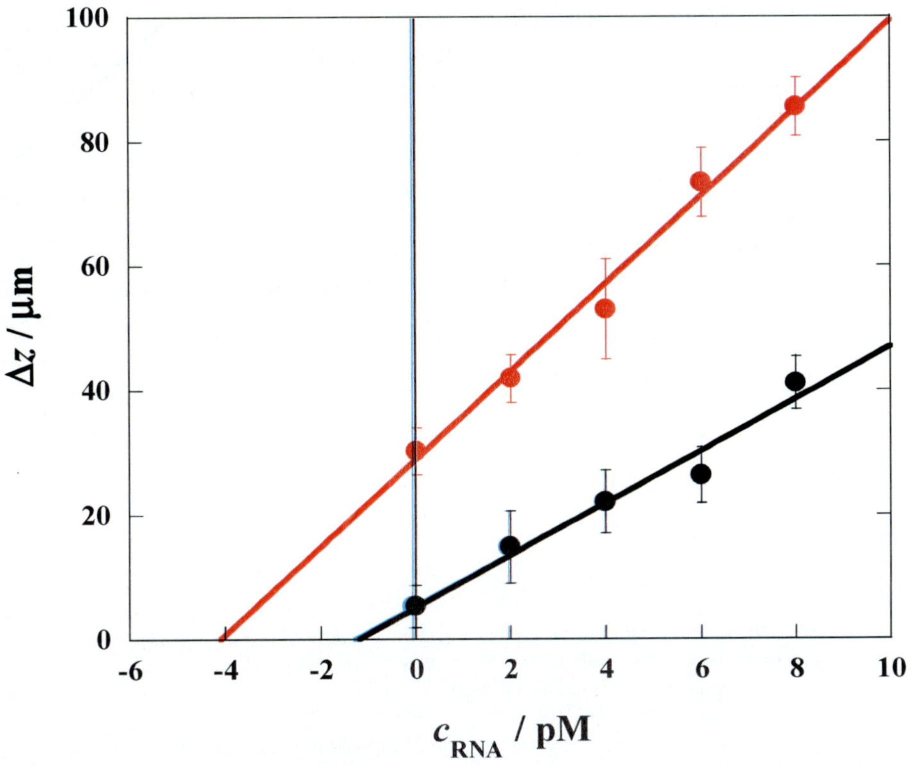

Fig. 5.16 Relationships between the spiked concentration of miR-21 or miR-122 (c_{RNA}) and Δz. Symbols: Black, miR-21 and red, miR-122. Reproduced from Ref. 7. Copyright 2018 American Chemical Society

The design of the direct interparticle hybridization between PMMA and AuNP was studied to evaluate the effect of the bp number and n_{PMMA} on sensitivity. The sensitivity decreased as the bp number and n_{PMMA} decreased, which was well explained by the calculations based on the established method. The largest difference in sensitivity was found between 8 and 9 bps. Although sensitivity became poor as n_{PMMA} decreased, selectivity was enhanced. Thus, n_{PMMA} is a useful parameter for adjusting the detection sensitivity and selectivity. Selectivity in the sandwich hybridization was predictable from that in direct interparticle hybridization. HIV-2 DNA was detected in the zmol order without modification to the target DNA.

An SNP detection scheme was designed by adjusting the affinity of the probe DNA to a target. SNP detection was successfully performed by adjusting the number of bp matching and n_{PMMA}. The dependence of Δz on n_{PMMA} was analyzed based on the equilibrium model. The theoretical estimation explained the experimental Δz values for any n_{PMMA} studied. The detection limit of wild DNA was 2700 at $n_{PMMA} = 2.3 \times 10^3$ mL^{-1}. This method was applicable to any SNP detection because sensitivity and selectivity could be controlled simply by adjusting n_{PMMA}.

The proposed approach was employed for the detection of multiple miRNAs in total RNA. PMMA microparticles with diameters of 9.57 and 6.33 μm were used for the detection of miR-21 and miR-122, respectively. These particles were distinguishable from a microscopic view. Annealing was necessary to prevent the undesirable hybridization of interferents. The quantification of miR-21 and miR-122 in total RNA solution was performed using the standard addition method. The concentrations of both miRNAs were in the pM order and were in good agreement with the reported values. However, the detection of three or more targets was difficult in this system. The use of particles with diameters larger than 10 μm leads to a decrease in spatial resolution, whereas particles with a diameter smaller than 4 μm cannot be entrapped by the CAG field. This problem may be solved by combining the present scheme with flow cytometry, which can recognize particles dyed with different fluorophores. The particle size effect was negligible in this case.

References

1. miRbase Web site (2018). https://www.mirbase.org/index.shtml (accessed April 24, 2018)
2. Miyagawa A, Harada M, Okada T (2018) Zeptomole biosensing of DNA with flexible selectivity based on acoustic levitation of a single microsphere binding gold nanoparticles by hybridization. ACS Sens 3:1870–1875
3. Owczarzy R, Moreira BG, You Y, Behlke MA, Walder JA (2008) Predicting stability of DNA duplexes in solutions containing magnesium and monovalent cations. Biochemistry 47:5336–5353
4. Fong LK, Wang Z, Schatz GC, Luijten E, Mirkin CA (2018) The role of structual enthalpy in spherical nucleic acid hybridization. J Am Chem Soc 140:6226–6230
5. Suzuki S, Komori M, Hirai M, Ureshino N, Kimura S (2012) Development of a novel, fully-automated genotyping system: principle and applications. Sensors 12:16614–16627
6. Jiang Y, Kimchi ET, Staveley-O'Carroll KF, Cheng H, Ajiani JA (2009) Assessment of K-ras mutation: a step toward personalized medicine for patients with colorectal cancer. Cancer 115:3609–3617
7. Miyagawa A, Harada M, Okada T (2018) Multiple MicroRNA quantification based on acoustic levitation of single microspheres after one-pot sandwich interparticle hybridizations. Anal Chem 900:13729–13735
8. Dong H, Lei J, Ding L, Ju H, Zhang X (2013) MicroRNA: function, detection, and bioanalysis. Chem Rev 113:6207–6233
9. Graybill RM, Bailey RC (2016) Emwrging biosensing approaches for microRNA analysis. Anal Chem 88:431–450
10. Chan HM, Chan LS, Wong RNS, Li HW (2010) Direct quantification of single-molecules of MicroRNA by total internal reflection fluorescence microscopy. Anal Chem 82:6911–6918
11. Feng C, Mao X, Shi H, Bo B, Chen X, Chen T, Zhu X, Li G (2017) Detection of microRNA: a point-of care testing method based on a pH-responsive and highly efficient isothermal amplification. Anal Chem 89:6631–6636

Chapter 6
Aptamer-Based Sensing of Small Organic Molecules

Abstract In this chapter, a system based on sandwich hybridization is applied to the detection of small organic molecules using an aptamer. Aptamer is a DNA molecule with a specific sequence that selectively binds to a target. Aptamer DNA mediates the binding between PMMA and AuNPs to form AuNP-bound PMMA. When the target interacts with the aptamer, AuNP is released from the PMMA, which induces the levitation coordinate shift. This concept is confirmed for several model targets in Sect. 6.2. The thermodynamics involved in this system is discussed in Sect. 6.3.

Keywords Aptamer · Dopamine · ATP · Ampicillin

6.1 Experimental

The probe and aptamer DNAs were purchased from Fasmac Co., Ltd. (Kanagawa, Japan). The adenosine triphosphate (ATP), dopamine (DA), and ampicillin (AMP) were selected as target molecules for aptamer-based sensing. The sequences of aptamer DNA molecules, which selectively bind to these target molecules, were determined according to the literature [1–3]. Table 6.1 lists the DNA sequences used in this section. Swine blood, including sodium citrate, was purchased from KAC Co. (Kyoto, Japan). This blood was 10^4 times diluted with water. The concentration of ATP was determined using the proposed detection scheme and a luciferase assay to validate the proposed method. The luminoassay was performed using an ATP determination kit (BA100) purchased from Toyo B-net (Tokyo, Japan). Luminescence was measured using a luminometer (Luminescensor Octa, ATTO, Japan).

AuNP-bound PMMA microparticles were prepared using the same method as described in Chaps. 4 and 5. Aptamer DNA mediated the binding between AuNP and PMMA particles by sandwich hybridization. The ratio of aptamer DNA to PMMA microparticles was kept constant at 10,000. In addition, the ratio of AuNPs to PMMA microparticles was kept constant at 11,000 for all experiments. Thus, the ratio of PMMA:aptamer:AuNP was 1:10,000:11,000.

FE-SEM images of PMMA were obtained using a JSM-7500F (JEOL, Japan). The FE-SEM was operated at 6 kV. High-resolution images were obtained using secondary electron detection. The samples were coated with Pt/Pd.

© The Author(s), under exclusive license to Springer Nature Singapore Pte Ltd. 2021

A. Miyagawa, *Acoustic Levitation-Based Trace-Level Biosensing*, Springer Theses,
https://doi.org/10.1007/978-981-16-1425-5_6

Table 6.1 Sequences of nucleotides used for the detection of ATP, AMP, and DA

Symbol	Sequence (5′ → 3′)	Functional group
ATP aptamer	CCTGGGGGAGTATTGCGGAGGAAGG	
ATP aptamer capture	ATACTCCCCCAGG-PMMA	Amino (3′)
ATP aptamer reporter	AuNP-CCTTCCTCCGCA	Amino (5′)
AMP aptamer	AAAGCGGGCGGTTGTATAGCGGAA	
AMP aptamer capture	ACCGCCCGCTTT-PMMA	Amino (3′)
AMP aptamer reporter	AuNP-TTCCGCTATACA	Amino (5′)
DA aptamer	GTCTCTGTGTGCGCCAGAGAACACTGG GGCAGATATGGGCCAGCACAGAATGAG GCCC	
DA aptamer capture	GGCGCACACAGAGAC-PMMA	Amino (3′)
DA aptamer reporter	AuNP-GGGCCTCATTCTGTG	Amino (5′)

6.2 Detection of Small Organic Molecules

Figure 6.1 shows a schematic representation of the principle for aptamer-based sensing [4]. AuNP-bound PMMA microparticles, where aptamers mediate the binding between PMMA and AuNP by complementary sandwich hybridization, were prepared. When the target molecules are added to the AuNP-bound PMMA microparticles, aptamer DNA interacts with the target molecule and then the dsDNA between the aptamer DNA and probe DNA is dissociated. The release of AuNPs from PMMA microparticles leads to a decrease in the density of the latter, which results in an increase in z.

ATP, DA, and AMP, used as small organic target molecules, are suitable model compounds because they have different binding constants with the aptamer molecules used in this study (6.00×10^{-6}, 1.00×10^{-7}, and 1.38×10^{-8} M^{-1}, respectively) [1–3]. ATP is important for the regulation of cellular metabolism [5–8]. The lack of ATP

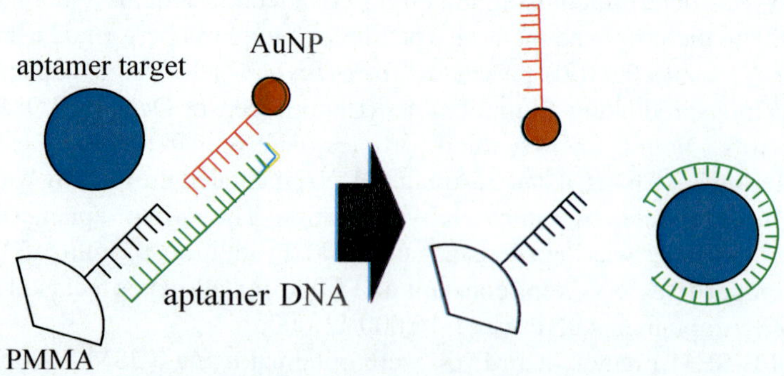

Fig. 6.1 Schematic representation of small molecules detection based on the sandwich hybridization of aptamer DNA. Reproduced from Ref. 4. Copyright 2020 American Chemical Society

is related to several diseases, such as angiocardiopathy, hypoglycemia, ischemia, and Parkinson's disease. DA is a neurotransmitter vital for memory, behavior, sleep, and cognition [9–11]. Low concentrations of DA are associated with Parkinson's disease, attention deficit hyperactivity disorder, and schizophrenia. Effective analyses of these compounds of physiological importance are required for biomedical monitoring. In addition, AMP is widely used to treat infectious bacterial diseases, such as pneumonia, gonorrhea, bronchitis, and venereal diseases, because of its broad spectrum and low cost [12, 13]. However, there are several side effects such as skin rashes, dizziness, and diarrhea. Thus, the development of effective sensing of AMP is also an important task.

The various concentrations of ATP, DA, and AMP were treated with individual aptamer-mediated AuNP-bound PMMA microparticles. Figure 6.2 shows the relationships between Δz and logarithmic concentrations of ATP, DA, and AMP (log c) [4]. Δz decreases as log c increases in a sigmoidal-shaped curve because of the release of AuNPs from the AuNP-bound PMMA. To determine the detection limit, the log c-Δz plots were converted into c-$\Delta z'$ plots; $\Delta z' = \Delta z_0 - \Delta z$, where Δz_0 is Δz of the AuNP-bound PMMA before the reaction with the target. These plots were

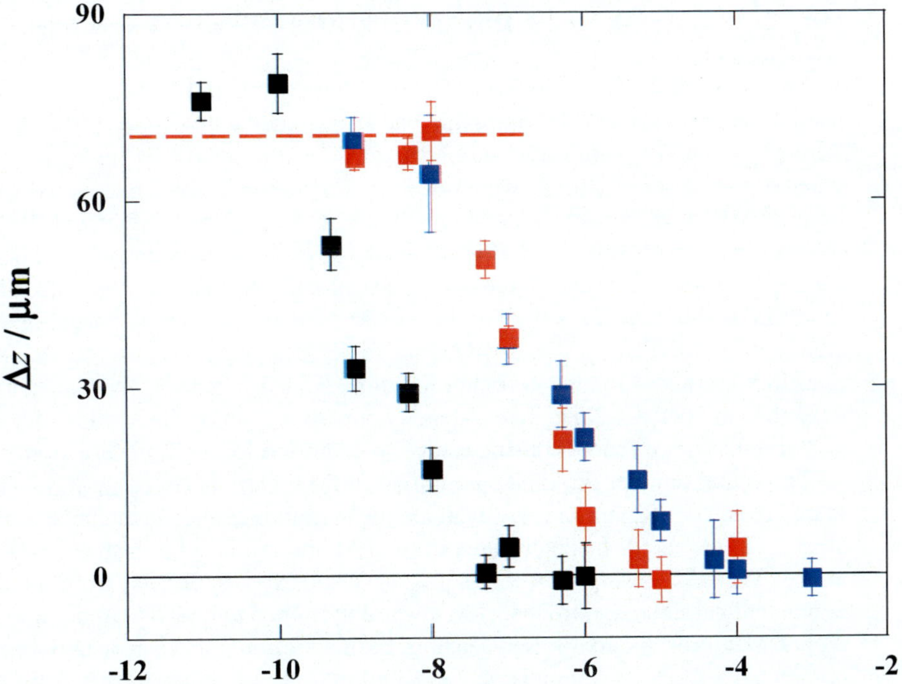

Fig. 6.2 Relationships between Δz of aptamer-mediated AuNP-bound PMMA and logarithmic concentration of target molecules: ATP (blue), DA (red), and AMP (black). Red dashed line represents the average maximum Δz at $r_{AuNP/PMMA} = 10{,}000$

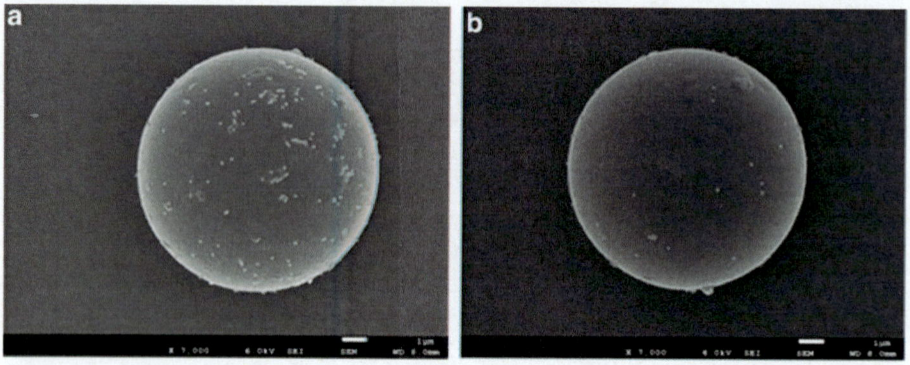

Fig. 6.3 FE-SEM images of an AuNP-bound PMMA microparticle used for ATP detection at **a** $c =$ 1 nM and **b** 1 mM. Reproduced from Ref. 4. Copyright 2020 American Chemical Society

approximated by the following logarithmic functions: $\Delta z' = -2.43 + 15.5 \log c$ for ATP, $\Delta z' = -11.7 + 20.5 \log c$ for DA, and $\Delta z' = 37.4 + 22.5 \log c$ for AMP. When the detection limit is defined as 3σ, it can be calculated from these equations; 3.7 nM ATP, 5.3 nM DA, and 42 pM AMP molecules were detectable, indicating that 1.2×10^8 ATP, 1.7×10^8 DA, and 1.3×10^6 AMP molecules per single PMMA are detectable.

Figure 6.3 shows the FE-SEM images of AuNP-bound PMMA microparticles used for ATP detection, [4] comparing particles treated with $c_{ATP} = 1$ nM and 1 mM. Almost no AuNPs were observed on the PMMA microparticles for $c_{ATP} = 1$ mM, whereas a number of AuNPs were observed for $c_{ATP} = 1$ nM. Thus, AuNPs remain on the PMMA microparticles when c is too low to remove AuNPs from PMMA by interaction with the aptamer, whereas the release of AuNP from PMMA is effectively caused by higher c. When $r_{AuNP/PMMA} = 10{,}000$, the surface coverage of AuNPs on the PMMA should be 25% if all of the AuNPs bind to the PMMA. From Fig. 6.3, the coverage of AuNP at 1 nM was calculated to be 3.8%, which is considerably smaller than that predicted for the complete binding. This arises from the same origin as described in Sect. 5.4.2, i.e., heterogeneity of carboxyl groups on the PMMA surface and elimination of AuNPs during sample preparation for SEM observation.

The actual number of AuNPs bound on PMMA particles ($n_{AuNP/PMMA}$) was estimated from Δz to quantitatively evaluate the thermodynamics in the present system, as $n_{AuNP/PMMA}$ cannot be determined from SEM observations for the reason discussed above. The relationship between $n_{AuNP/PMMA}$ and Δz was derived from the device-dependent parameter, a. In Sect. 3.5, a was determined to be 0.040 under this instrumental setup for aptamer-based sensing. In this section, all Δz measurements were conducted at $V = 7.5$ V, that is, $E_{ac} = 2.25$ J m^{-3}. The relationship between Δz and $n_{AuNP/PMMA}$ at $V = 7.5$ V was calculated to determine $n_{AuNP/PMMA}$ from Δz (Fig. 6.4) [4]. A linear relationship was confirmed in the range $n_{AuNP/PMMA} = 0$–6000:

$$\Delta z = 1.48 \times 10^{-2} n_{AuNP/PMMA} \tag{6.1}$$

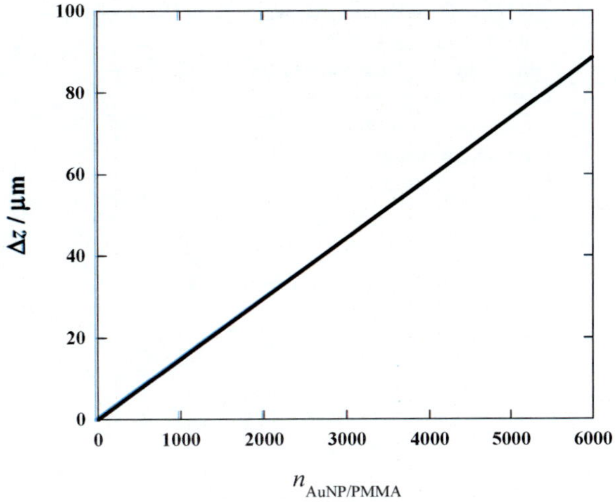

Fig. 6.4 Calculated Δz with $n_{\mathrm{AuNP/PMMA}}$ at $E_{\mathrm{ac}} = 2.25\ \mathrm{J\ m^{-3}}$. $a = 0.040$. Reproduced from Ref. 4. Copyright 2020 American Chemical Society

The sandwich hybridization system provided $\Delta z = 56\ \mu\mathrm{m}$ at $r_{\mathrm{RNA/PMMA}} = 10{,}000$ and $r_{\mathrm{AuNP/PMMA}} = 11{,}000$, as shown in Sect. 5.4. Substituting the Δz value into Eq. (6.1) results in $n_{\mathrm{AuNP/PMMA}} = 3783$, which indicates that the reaction ratio of AuNPs with PMMA is 34%. This reaction ratio is considerably larger than that determined from FE-SEM images (6.8%, Fig. 5.12). This may support the reasons indicated above for the low coverage of SEM images. When $r_{\mathrm{AuNP/PMMA}} = 10{,}000$, $\Delta z = 71 \pm 5.2\ \mu\mathrm{m}$, as shown in Fig. 6.2. Substituting this value in Eq. (6.1) provides $n_{\mathrm{AuNP/PMMA}} = 4736 \pm 338$, which indicates that the surface coverage of AuNPs is 12% at $r_{\mathrm{AuNP/PMMA}} = 10{,}000$. Thus, the reaction ratio was approximately 50% in this case. This value is larger than that obtained in the studies of miRNA. This is attributed to the difference in the bp number involved in the interparticle hybridization, i.e., 22 bp for miRNA and 25 bp for ATP aptamer.

This detection scheme was applied to the determination of ATP in swine blood. The blood was 10^4 times diluted with water. The known concentration of ATP was spiked into the diluted blood. After the reaction of AuNP-bound PMMA microparticles ($r_{\mathrm{AuNP/PMMA}} = 10{,}000$) with ATP in the diluted blood and spiked blood samples, Δz for the treated PMMA was measured in the CAG field. Figure 6.5 shows the relationships between log c and $\Delta z/\Delta z_0$, where Δz_0 represents Δz at the plateau region with lower c [4]. In Fig. 6.5, black and red plots represent the results measured in standard solutions prepared in water and swine blood, respectively. Almost linear relationships were confirmed, and two plots provided almost identical traces, indicating that the blood matrix did not interfere with the reactions involved in this detection scheme.

The concentration of ATP in the diluted blood was determined to be 17.6 nM, i.e., in the original blood it was 176 μM. The ATP concentration was validated by a luminescence assay, which provided an ATP concentration of 202 μM. This value agrees with the result in the CAG field. Thus, the proposed method is applicable to real samples.

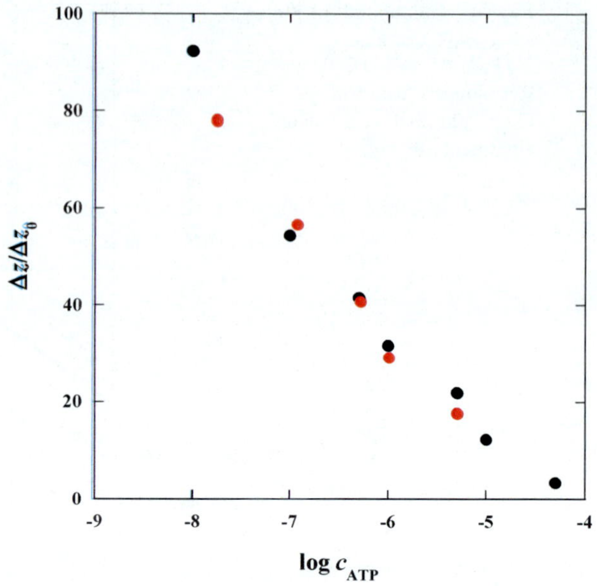

Fig. 6.5 Calibration graphs for determining ATP in swine blood. Black and red plots represent $\Delta z/\Delta z_0$ for the standard solution prepared in water and swine blood, respectively

6.3 Equilibrium Analysis in the Aptamer-Based Sensing System

Figure 6.1 suggests that Δz is determined by the balance between the stability of the aptamer-target complex and the strength of the aptamer and probe DNA duplex formation. The equilibrium model in the present system is depicted in Fig. 6.6 a for the capture DNA anchored on PMMA, B for the aptamer DNA, C for the reporter DNA-anchored AuNP, D for AuNPs bound to PMMA, E for the target molecule, and F for the aptamer-target complex. Fluorescence measurements using FITC-labeled complementary DNA indicated that 1.6×10^7 capture DNA molecules were anchored on a PMMA particle. All capture DNAs were assumed to be involved in the hybridization with aptamer DNA. In contrast, only one reporter DNA molecule on an AuNP was bound to the aptamer DNA trapped on the PMMA. Once one reporter DNA molecule binds to the PMMA, other reporter DNA molecules on the AuNP cannot be involved in hybridization. Therefore, the concentration of C (Fig. 6.6) should be equal to the concentration of AuNPs rather than the concentration of the reporter DNA in the reaction system. Figure 6.7 shows the relationship between log K_1 and $n_{\mathrm{AuNP/PMMA}}$ at $r_{\mathrm{AuNP/PMMA}} = 10{,}000$ [4]. Because $n_{\mathrm{AuNP/PMMA}}$ was determined to be 4736 ± 338 from Fig. 6.4 as already described, K_1 was determined to be 1.58×10^{20} M^{-2}.

Typically, the deviation of Δz is 3–5 μm in this system. The theoretical detection limit calculated from Eq. (6.1) was $n_{\mathrm{AuNP/PMMA}} = 202$–$337$. Thus, from Fig. 6.7, the levitation coordinate shift is detectable if log $K_1 > 18.5$. However, it should be noted that the relationship shown in Fig. 6.7 does not consider the individual association constants between A and B ($K_{\mathrm{A\text{-}B}}$), and between the A-B complex and C ($K_{\mathrm{AB\text{-}C}}$).

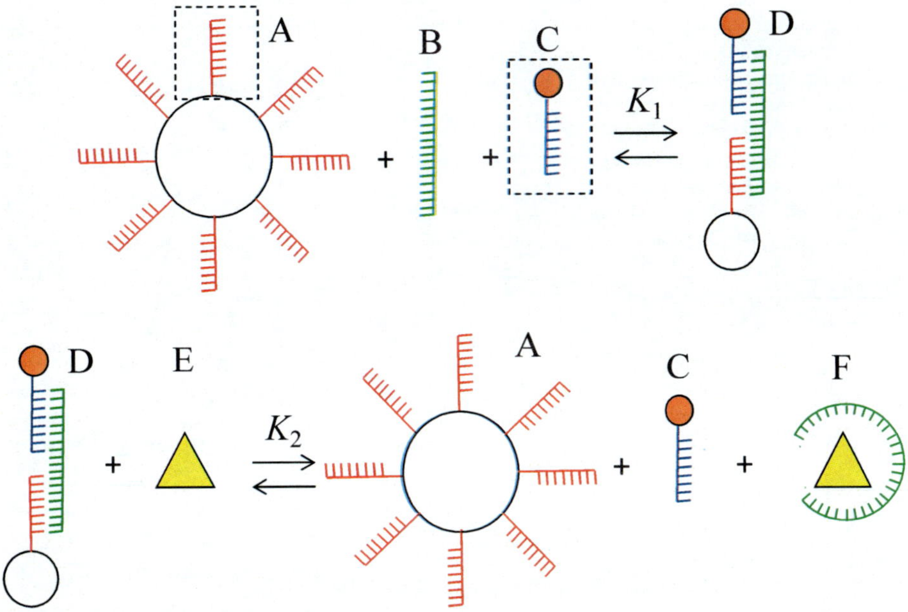

Fig. 6.6 Equilibrium model in the present study

Fig. 6.7 Calculated relationship between log K_1 and $n_{\mathrm{AuNP/PMMA}}$ at $r_{\mathrm{AuNP/PMMA}} = 10{,}000$. Reproduced from Ref. 4. Copyright 2020 American Chemical Society

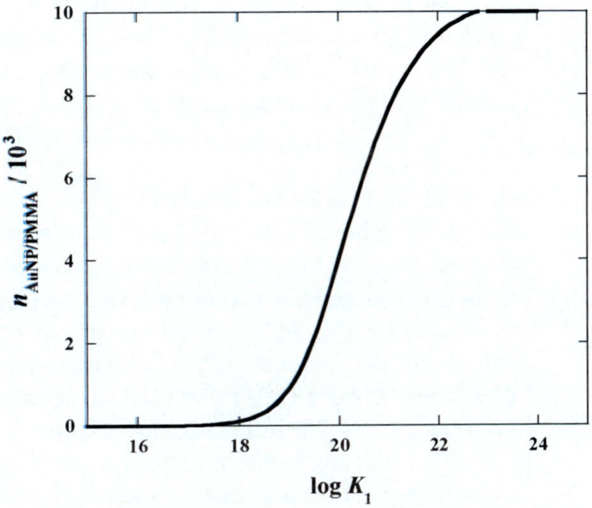

Thus, this relationship is not valid for the cases where a large difference is involved between K_{AB} and $K_{\mathrm{AB\text{-}C}}$, for example, log $K_{\mathrm{A\text{-}B}} = 10^{22}$ and $K_{\mathrm{AB\text{-}C}} = 10$. However, in this system, the A-B and B-C dsDNA are composed of almost the same length, indicating that we can predict the stability of AuNP-bound PMMA based on the model shown in Fig. 6.7.

Figure 6.8 shows the relationships between log c and $\Delta z/\Delta z_0$ obtained for the

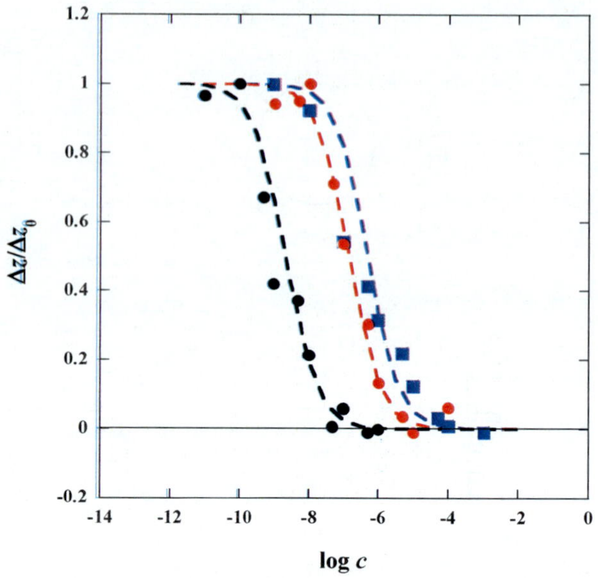

Fig. 6.8 Relationships between $\Delta z/\Delta z_0$ and logarithmic concentration of target molecules for ATP (blue), DA (red), and AMP (black). The dashed curves were calculated assuming $K_1 = 1.58 \times 10^{-20}$ M^{-2}, and $K_2 = 1.00 \times 10^{-11}$, 1.58×10^{-13}, and 1.58×10^{-14} M for AMP, DA, and ATP, respectively

three aptamer-targets [4]. The sigmoidal curves were simulated assuming a common K_1 value ($= 1.58 \times 10^{20}$ M^{-2}) and $K_2 = 1.00 \times 10^{-11}$, 1.58×10^{-13}, and 1.58×10^{-14} M for AMP, DA, and ATP, respectively. The dissociation constant of the aptamer-target complex was given by $K_3 = 1/(K_1K_2)$. Thus, K_3 values for AMP, DA, and ATP were 6.33×10^{-10}, 4.00×10^{-8}, and 1.59×10^{-7} M^{-1}, respectively. These values are smaller than the values reported in the literature (1.38×10^{-8}, 1.00×10^{-7}, and 6.00×10^{-6} M^{-1}, respectively) [1–3]. The dsDNA formation and the binding affinity of the aptamer-target complex are influenced by the buffer salt concentration and/or composition [14–17]. For example, Baldrich et al. studied the structural instability of thrombin-binding aptamer (TBA) in the presence of different types of salt [13]. When KCl was added, TBA-thrombin complex was stable, whereas the addition of Na^+ or Mg^{2+} caused the disruption of the complex. In contrast, Munzar et al. reported the enhancement of the hybridization affinity of the aptamer-complementary DNA duplex with the increase in the concentration of ions in a buffer [16]. Thus, the interactions between DNAs and between aptamer and target are complexly affected by various factors. Therefore, the dissociation constants of a target-aptamer complex determined under different conditions cannot be simply compared. In Fig. 6.6, the 1:1 stoichiometry in the aptamer-target reactions was assumed. However, 1:2 complexation of the aptamer with ATP molecules was also reported [18]. Figure 6.9 shows the curve of the $\Delta z/\Delta z_0$ calculated for ATP assuming 1:2 complexation. The calculated curve is not consistent with the experimental values. A better fit is seen in Fig. 6.8, where a 1:1 complexation was assumed. Thus, the 1:1 complexation is reasonable for this system. Hence, the present method allows to evaluate the aptamer–target complex interactions in various media. Because no structural modification or labeling

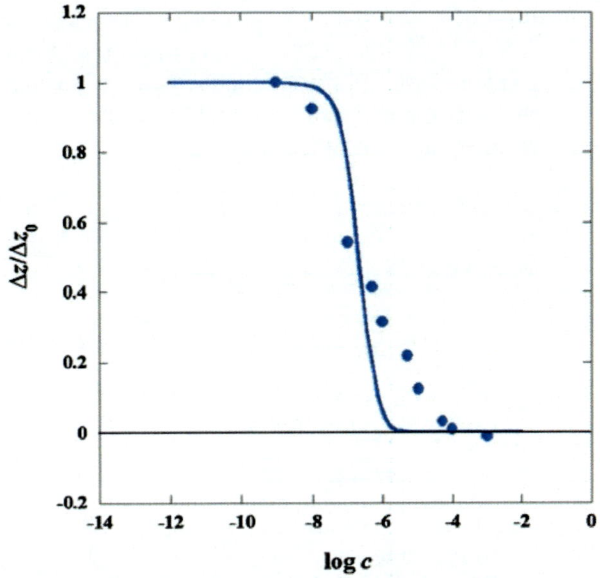

Fig. 6.9 Relationships between $\Delta z/\Delta z_0$ and logarithmic concentration of target molecules for ATP assuming the 1:2 complexation. The curve represents the result of calculation with $K_1 = 1.58 \times 10^{20}\ M^{-2}$ and $K_2 = 1.00 \times 10^{-7}$ M

of a target or aptamer is required in the proposed method, it can be a powerful tool for the screening of useful aptamer molecules.

6.4 Conclusion

The aptamer-based sensing was designed using the CAG field, and small organic molecules such as ATP, DA, and AMP were successfully detected. The dissociation of AuNP-bound PMMA due to the interaction of the aptamer with a target was utilized in the detection scheme, which led to a decrease in Δz. As c increased, Δz decreased in the c-range depending on the aptamer-target binding constant. The detection limits of ATP, DA, and AMP were 3.7 nM, 5.3 nM, and 42 pM, respectively. This approach was applied to quantify ATP in swine blood. The concentration (176 μM) determined by the proposed method was in good agreement with that determined by the luciferase assay. Thus, the application to real samples was successfully demonstrated.

The reaction of the aptamer with the target was thermodynamically analyzed based on the equilibrium model shown in Fig. 6.7. The equilibrium constant for the formation of AuNP-bound PMMA, K_1, was calculated to be $1.58 \times 10^{-20}\ M^{-2}$ from the number of AuNPs bound to the PMMA microparticles. The dissociation constants for the aptamer–target complex were determined to be 6.33×10^{-10} for AMP, 4.00 $\times 10^{-8}$ for DA, and $1.59 \times 10^{-7}\ M^{-1}$ for ATP by the analyses of log c-$\Delta z/\Delta z_0$ curves. This indicates that the proposed method is applicable to the thermodynamic evaluation of aptamer–target interactions.

Various aptamer molecules have been developed for inorganic ions (K^+, Hg^{2+}, etc.), organic molecules (cocaine, ibuprofen, etc.), large biomolecules (peptides and proteins), and microorganisms (bacteria and cells). The present approach can be extended to other targets without significant modification because the probe DNA sequences are simply adaptable to that of an appropriate aptamer sequence. I believe that this method can be widely used in a variety of fields such as environmental, medical, and biological sciences.

References

1. Huizenga DE, Szostak JW (1995) A DNA aptamer that binds adenosine and ATP. Biochemistry 34:656–665
2. Song KM, Jeong E, Jeon W, Cho M, Ban C (2012) Aptasensor for ampicillin using gold nanoparticle based dual fluorescence-colorimetric methods. Anal Bioanal Chem 402:2153–2161
3. Walsh R, DeRosa MC (2009) Retention of function in the DNA homolog of the RNA dopamine aptamer. Biochem Bioph Res Co 388:732–735
4. Miyagwa A, Okada Y, Okada T (2020) Aptamer-based sensing of small organic molecules by measuring levitation coordinate of single microsphere in combined acoustic-gravitational field. ACS Omega 5:3542–3549
5. Ma C, Lin C, Wang Y, Chen X (2016) DNA-based ATP sensing. TrAC-Trend Anal Chem 77:226–241
6. Rajendran M, Dane E, Conley J, Tantama M (2016) Imaging Adenosine Triphosphate (ATP). Biol Bull 231:73–84
7. Dong J, Zhao M (2016) *In-vivo* fluorescence imaging of adenosine 5'-triphosphate. TrAC-Trend Anal Chem 80:190–203
8. Ng S, Lim HS, Ma Q, Gao Z (2016) Optical aptasensors for adenosine triphosphate. Theranostics 6:1683–1702
9. Sajid M, Nazal MK, Mansha M, Alsharaa A, Jillani SMS, Basheer C (2016) Chemically modified electrodes for electrochemical detection of dopamine in the presence of uric acid and ascorbic acid: a review. TrAC-Trend Anal Chem 76:15–29
10. Pandikumar A, How GTS, See TP, Omar FS, Jayabal S, Kamali KZ, Yusoff N, Jamil A, Ramaraj R, John SA, Lim HN, Huang NM (2014) Graphene and its nanocomposite material based electrochemical sensor platform for dopamine. RSC Adv 4:63296–63323
11. Rasheed PA, Lee JS (2017) Recent advances in optical detection of dopamine using nanomaterials. Microchim Acta 184:1239–1266
12. Mehlhorn A, Rahimi P, Joseph Y (2018) Aptamer-based biosensors for antibiotic detection: a review. Biosensors 8:54
13. Shrivas K, Sahu J, Maji P, Sinha D (2017) Label-free selective detection of ampicillin drug human urine samples using silver nanoparticles as a colorimetric sensing probe. New J Chem 41:6685–6692
14. Baldrich E, Restrepo A, O'Sullivan CK (2004) Aptasensor development: elucidation of critical parameters for optimal aptamer performance. Anal Chem 76:7053–7063
15. Haq I, Lincoln P, Suh D, Norden B, Chowdhry BZ, Chaires JB (1995) Interaction of Δ- and Λ-$[Ru(phen)_2DPPZ]^{2+}$ with DNA: a calorimetric and equilibrium binding study. J Am Chem Soc 117:4788–4796
16. Munzar JD, Ng A, Juncker D (2019) Duplex aptamers: history, design, theory, and application to biosensing. Chem Soc Rev 48:1390–1419
17. Wang J, Jiang Y, Zhou C, Fang X (2005) Aptamer-based ATP assay using a luminescent light switching complex. Anal Chem 77:3542–3546
18. Jhaveri SD, Kirby R, Conrad R, Maglott EJ, Bowser M, Kennedy RT, Glick G, Ellington AD (2000) Designed signaling aptamers that transduce molecular recognition to changes in fluorescence intensity. J Am Chem Soc 122:2469–2473

Chapter 7
Conclusion and Outlook

Acoustic levitation-based biosensing at the zmol level was demonstrated in this thesis. This method is versatile and can be tailored to detect a variety of targets, such as proteins, coenzymes, nucleotides, and organic molecules. The principle discussed in this thesis is based on measurements of a levitation coordinate shift induced by the density change due to AuNP binding. The levitation behaviors of gold-plated PMMA and AuNP-bound PS obeyed the prediction using Eq. (2.70). The linear relationships between Δz and $n_{\mathrm{AuNP/MP}}$ for microparticles (EP, PS, and PMMA) were confirmed. These results indicate that the surface reaction of microparticles is detectable in the CAG field as a density change is induced by AuNP binding. As described above, the CAG field can convert subtle density changes into measurable levitation coordinate shifts. Thus, highly sensitive detection at the zmol order is feasible without signal amplification.

This principle was used for kinetic and thermodynamic evaluations of the reactions on the particle surface. The kinetic analysis showed that the equilibrium constant of the avidin–biotin reaction on the particle was smaller than that in solution because of the decrease in the apparent diffusion coefficient. In addition, the thermodynamics for DNA hybridization and aptamer–target complexation were quantitatively evaluated. The association constants for direct interparticle hybridization were different from those predicted from ΔG provided by the NN model because of two opposite effects: the enthalpic effect on the DNA structure and the decrease in molecular diffusion by anchoring DNA on particles. In contrast, the association constants for sandwich interparticle hybridization were well explained by theoretical ΔG because the target DNAs were free and not anchored on the particle. The dissociation constants of the aptamer-target complex for ATP, DA, and AMP were also determined under the present conditions. These results show the high potential of the present concept not only for zmol biosensing but also for kinetic and thermodynamic analyses.

This thesis demonstrates that the present approach is applicable to the detection of various targets because the detection system can be modified simply by replacing the reaction. For example, the aptamer-based sensing scheme can be extended to any

© The Author(s), under exclusive license to Springer Nature Singapore Pte Ltd. 2021

A. Miyagawa, *Acoustic Levitation-Based Trace-Level Biosensing*, Springer Theses,
https://doi.org/10.1007/978-981-16-1425-5_7

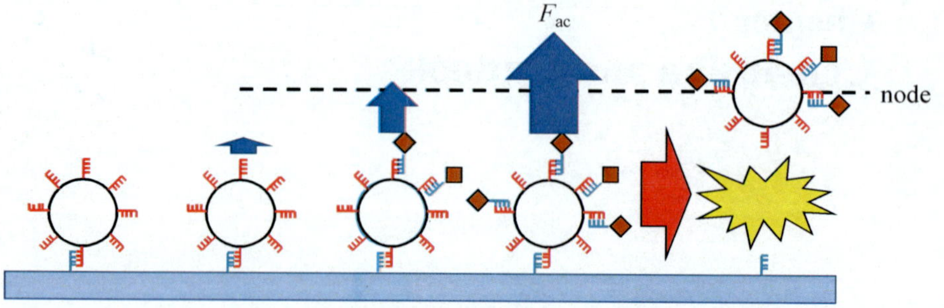

Fig. 7.1 Schematic representation of a novel detection scheme for weak binding

target such as inorganic ions, organic molecules, large biomolecules, and microorganisms, if an appropriate aptamer is available. Various SNPs are also detectable by adjusting the bpnumber and DNA sequence. Thus, this approach is useful in various fields such as biochemistry, environmental, diagnosis, and forensic sciences.

The detection of reactions on the microparticle surface requires only a low sample volume because the microparticle has a very small reaction space. Although the levitation coordinate shift of a single microparticle was evaluated in the present study, the reaction was conducted in bulk solutions containing a large number of microparticles and AuNPs. This problem can be solved by inducing an in situ reaction in an observation of microcell. It also allows to detect the reaction with a small equilibrium constant that cannot be evaluated by the present approach and also to evaluate the in situ reaction dynamics.

Figure 5.3 indicates that an interparticle reaction with an association constant of $K > 10^8$ is required for the present scheme. A novel detection scheme applicable to the reactions of a smaller K should be designed to overcome this limitation. Figure 7.1 proposes a novel concept. The microparticles are fixed on a substrate through weak binding. When an ultrasound standing wave field is formed in this system, the particle is pulled toward the node of the standing wave. The particle remains on the substrate while the binding force (F_b) is larger than F_{ac}. However, F_{ac} is a function of the density (Eqs. (2.63–2.64) and increases as AuNP binding proceeds. When $F_{ac} > F_b$, the movement of the particle to the node is induced. The detection is performed by statistical evaluation based on the number of microparticles moving to the node. This approach is suitable for the detection of weak interactions because it is based on the breaking of the binding between the microparticle and the substrate.

The trace analysis based on the drastic change in A in Eq. (2.70) is also possible. Figure 7.2 shows the principle of this scenario. As described in Sect. 1.3, the sign of A determines the direction of particle movement. If $A > 0$, the particle moves toward the node of the standing wave, as discussed in this thesis. In contrast, a particle moves to the antinode when $A < 0$. This suggests that a change in the sign of A induces a drastic translocation of the levitation position. Nanobubbles (NB) and hollow particles have low densities and high compressibility. The binding of these

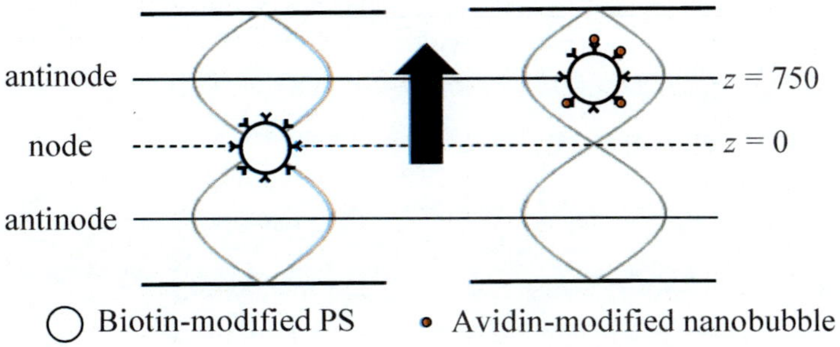

Fig. 7.2 Schematic representation of the detection based on the change in *A*

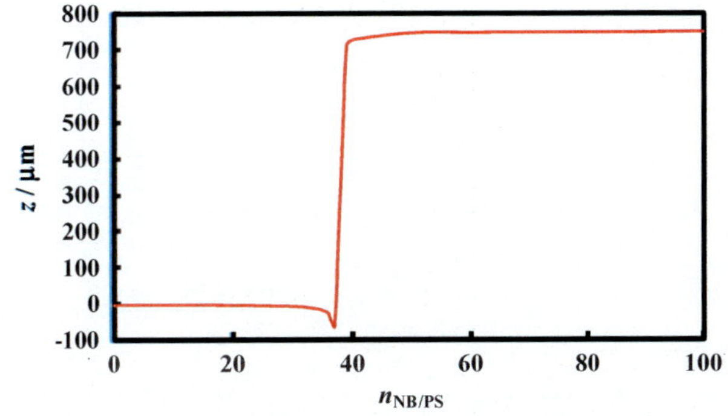

Fig. 7.3 Relationship between z and $n_{NB/PS}$ calculated assuming that a 10 μm PS particle reacts with 100 nm NBs

low-density nanoparticles to a microparticle like a PS particle induces the positive-to-negative switching of *A* and the drastic change in the levitation coordinates of the microparticle. This levitation coordinate change was simulated assuming that a 10 μm PS particle reacts with 100 nm NBs ($\rho_{NB} = 1$ g cm^{-3} and $\gamma_{NB} = 8.62 \times 10^{-6}$ Pa^{-1}). Figure 7.3 shows the relationship between z and the number of NBs bound to a PS ($n_{NB/PS}$). z drastically increases at $n_{NB/PS} = 38$, indicating that the sign of *A* changes from a positive to a negative value. The threshold $n_{NB/PS}$ is easily detectable, and a small coordinate shift does not need to be measured. The threshold value is controlled by changing the number of molecules introduced to both PS and NB. This approach should be more sensitive than the method proposed in this thesis.

This thesis demonstrated the feasibility of trace-level biosensing using the CAG field. I believe that these concepts are useful for trace analyses in important fields such as analytical chemistry, biochemistry, environmental science, diagnosis, and forensic science and also an efficient tool for evaluating the kinetics and thermodynamics of reactions that are difficult to probe using other well-established methods. I expect the proposed approach to be more extensively utilized as an efficient tool.

Curriculum Vitae

Name: Akihisa Miyagawa

Affiliation: Division of Chemistry, Faculty of Pure and Applied Sciences, University of Tsukuba

Address: 1-1-1, Tennodai, Tsukuba, Ibaraki, 305-8577, Japan

E-mail: miyagawa.akihisa.gf@u.tsukuba.ac.jp

Employments

- Assistant Professor, Faculty of Pure and Applied Sciences, University of Tsukuba (April 2020-)
- JSPS research fellow, Tokyo Institute of Technology (April 2019-March 2020)

Education

- Ph.D., Department of Chemistry, Tokyo Institute of Technology, March 2020
- M.Sc., Department of Chemistry, Tokyo Institute of Technology, March 2017
- B. Sc., Department of Applied Chemistry, Meiji University, March 2015

Research Interest

Analytical Chemisty, Molecular Spectroscopy, Chemistry under the Hydrostatic Pressure, and Interfacial Chemistry.

© The Editor(s) (if applicable) and The Author(s), under exclusive license to Springer Nature Singapore Pte Ltd. 2021

A. Miyagawa, *Acoustic Levitation-Based Trace-Level Biosensing*, Springer Theses,
https://doi.org/10.1007/978-981-16-1425-5

BOOK-1

01L

HB

Lecture Notes in Electrical Engineering

Volume 741

Series Editors

Leopoldo Angrisani, Department of Electrical and Information Technologies Engineering, University of Napoli Federico II, Naples, Italy
Marco Arteaga, Departament de Control y Robótica, Universidad Nacional Autónoma de México, Coyoacán, Mexico
Bijaya Ketan Panigrahi, Electrical Engineering, Indian Institute of Technology Delhi, New Delhi, Delhi, India
Samarjit Chakraborty, Fakultät für Elektrotechnik und Informationstechnik, TU München, Munich, Germany
Jiming Chen, Zhejiang University, Hangzhou, Zhejiang, China
Shanben Chen, Materials Science and Engineering, Shanghai Jiao Tong University, Shanghai, China
Tan Kay Chen, Department of Electrical and Computer Engineering, National University of Singapore, Singapore, Singapore
Rüdiger Dillmann, Humanoids and Intelligent Systems Laboratory, Karlsruhe Institute for Technology, Karlsruhe, Germany
Haibin Duan, Beijing University of Aeronautics and Astronautics, Beijing, China
Gianluigi Ferrari, Università di Parma, Parma, Italy
Manuel Ferre, Centre for Automation and Robotics CAR (UPM-CSIC), Universidad Politécnica de Madrid, Madrid, Spain
Sandra Hirche, Department of Electrical Engineering and Information Science, Technische Universität München, Munich, Germany
Faryar Jabbari, Department of Mechanical and Aerospace Engineering, University of California, Irvine, CA, USA
Limin Jia, State Key Laboratory of Rail Traffic Control and Safety, Beijing Jiaotong University, Beijing, China
Janusz Kacprzyk, Systems Research Institute, Polish Academy of Sciences, Warsaw, Poland
Alaa Khamis, German University in Egypt El Tagamoa El Khames, New Cairo City, Egypt
Torsten Kroeger, Stanford University, Stanford, CA, USA
Qilian Liang, Department of Electrical Engineering, University of Texas at Arlington, Arlington, TX, USA
Ferran Martín, Departament d'Enginyeria Electrònica, Universitat Autònoma de Barcelona, Bellaterra, Barcelona, Spain
Tan Cher Ming, College of Engineering, Nanyang Technological University, Singapore, Singapore
Wolfgang Minker, Institute of Information Technology, University of Ulm, Ulm, Germany
Pradeep Misra, Department of Electrical Engineering, Wright State University, Dayton, OH, USA
Sebastian Möller, Quality and Usability Laboratory, TU Berlin, Berlin, Germany
Subhas Mukhopadhyay, School of Engineering & Advanced Technology, Massey University, Palmerston North, Manawatu-Wanganui, New Zealand
Cun-Zheng Ning, Electrical Engineering, Arizona State University, Tempe, AZ, USA
Toyoaki Nishida, Graduate School of Informatics, Kyoto University, Kyoto, Japan
Federica Pascucci, Dipartimento di Ingegneria, Università degli Studi "Roma Tre", Rome, Italy
Yong Qin, State Key Laboratory of Rail Traffic Control and Safety, Beijing Jiaotong University, Beijing, China
Gan Woon Seng, School of Electrical & Electronic Engineering, Nanyang Technological University, Singapore, Singapore
Joachim Speidel, Institute of Telecommunications, Universität Stuttgart, Stuttgart, Germany
Germano Veiga, Campus da FEUP, INESC Porto, Porto, Portugal
Haitao Wu, Academy of Opto-electronics, Chinese Academy of Sciences, Beijing, China
Junjie James Zhang, Charlotte, NC, USA

The book series *Lecture Notes in Electrical Engineering* (LNEE) publishes the latest developments in Electrical Engineering - quickly, informally and in high quality. While original research reported in proceedings and monographs has traditionally formed the core of LNEE, we also encourage authors to submit books devoted to supporting student education and professional training in the various fields and applications areas of electrical engineering. The series cover classical and emerging topics concerning:

- Communication Engineering, Information Theory and Networks
- Electronics Engineering and Microelectronics
- Signal, Image and Speech Processing
- Wireless and Mobile Communication
- Circuits and Systems
- Energy Systems, Power Electronics and Electrical Machines
- Electro-optical Engineering
- Instrumentation Engineering
- Avionics Engineering
- Control Systems
- Internet-of-Things and Cybersecurity
- Biomedical Devices, MEMS and NEMS

For general information about this book series, comments or suggestions, please contact leontina.dicecco@springer.com.

To submit a proposal or request further information, please contact the Publishing Editor in your country:

China

Jasmine Dou, Editor (jasmine.dou@springer.com)

India, Japan, Rest of Asia

Swati Meherishi, Editorial Director (Swati.Meherishi@springer.com)

Southeast Asia, Australia, New Zealand

Ramesh Nath Premnath, Editor (ramesh.premnath@springernature.com)

USA, Canada:

Michael Luby, Senior Editor (michael.luby@springer.com)

All other Countries:

Leontina Di Cecco, Senior Editor (leontina.dicecco@springer.com)

**** This series is indexed by EI Compendex and Scopus databases. ****

More information about this series at http://www.springer.com/series/7818

Zahriladha Zakaria · Seyed Sattar Emamian
Editors

Advances in Electrical and Electronic Engineering and Computer Science

Springer

Editors
Zahriladha Zakaria
Faculty of Electronic Engineering
and Computer Engineering
Centre for Telecommunication
Research and Innovation
Universiti Teknikal Malaysia Melaka
Melaka, Malaysia

Seyed Sattar Emamian
Sigma Research and Consulting
Delft, Zuid-Holland, The Netherlands

ISSN 1876-1100 ISSN 1876-1119 (electronic)
Lecture Notes in Electrical Engineering
ISBN 978-981-33-6489-9 ISBN 978-981-33-6490-5 (eBook)
https://doi.org/10.1007/978-981-33-6490-5

© The Editor(s) (if applicable) and The Author(s), under exclusive license to Springer Nature Singapore Pte Ltd. 2021
This work is subject to copyright. All rights are solely and exclusively licensed by the Publisher, whether the whole or part of the material is concerned, specifically the rights of translation, reprinting, reuse of illustrations, recitation, broadcasting, reproduction on microfilms or in any other physical way, and transmission or information storage and retrieval, electronic adaptation, computer software, or by similar or dissimilar methodology now known or hereafter developed.
The use of general descriptive names, registered names, trademarks, service marks, etc. in this publication does not imply, even in the absence of a specific statement, that such names are exempt from the relevant protective laws and regulations and therefore free for general use.
The publisher, the authors and the editors are safe to assume that the advice and information in this book are believed to be true and accurate at the date of publication. Neither the publisher nor the authors or the editors give a warranty, expressed or implied, with respect to the material contained herein or for any errors or omissions that may have been made. The publisher remains neutral with regard to jurisdictional claims in published maps and institutional affiliations.

This Springer imprint is published by the registered company Springer Nature Singapore Pte Ltd.
The registered company address is: 152 Beach Road, #21-01/04 Gateway East, Singapore 189721, Singapore

Preface

This book presents a compilation of research works covering the fields of computer science, electrical and electronic engineering. All manuscripts in this volume were presented during the 2nd International Conference on Computer Science, Electrical and Electronic Engineering 2020 (ICCEE 2020) which was conducted through a virtual presentation on 18 August 2020. This conference holds a vital role in as a catalyst in seeking wisdom, sharing thoughts and opinions to promote better understanding of engineering and information technology and to strive for quality research towards Fourth Industrial Revolution (IR 4.0).

The editor(s) of the proceeding would like to express the utmost gratitude and thanks to all reviewers in the technical team for making this volume a success.

Delft, The Netherlands — Dr. Seyed Sattar Emamian
Melaka, Malaysia — Prof. Dr. Zahriladha Zakaria

Acknowledgments

The editors would like to thank all the members of the local organizing committee who helped organize the 2nd International Conference on Computer Science, Electrical and Electronic Engineering 2020 (ICCEE 2020), which was conducted through a virtual presentation on 18 August 2020. We would like to thank the colleagues and staff members at the institutions and organizations that served as partners for the international conference. Their support in organizing a successful conference has helped the editors to gather ideas and papers presented in this book. The editors are grateful to all the speakers who attended the conference and shared from their wealth of experience some exciting findings who have further propelled us to publish this book.

The editors also appreciate various people, including the production team at Springer, who helped and contributed to the creation of this book. We thank all the authors and contributors who presented at the conference and sent us their papers for peer-review. The editors would like to thank and appreciate the peer-reviewers for their suggestions, comments, efforts and time spent to go over all the papers. The creation of this book has helped us to become a formidable team. The process has been enjoyable, challenging, inspiring and more peaceful than we ever thought. We thank you all!

Delft, The Netherlands
Melaka, Malaysia

Dr. Seyed Sattar Emamian
Prof. Dr. Zahriladha Zakaria

Contents

A Mobile Augmented Reality Interface on Additive Manufacturing

Yih Bing Chu and Chee Wah Chang

Abstract The increasing customized and complexity of products makes the handling of additive manufacturing becomes more sophisticated. Amidst the growing demands of personalized product, interpretation of input from customer to the specification of product is now a challenge for the industry. In view of this, augmented reality (AR), a technology which projects virtual content of target object on real world scenery can be used to interpret the design information and project the detail onto display accordingly. Through visual interaction, the technical information on additive manufacturing can be easily visualised and flow of the process can be managed and planned on preliminary before actual production. This helps prevent the misunderstanding of information conveyed through verbal communication and subsequently satisfactory of the product fabricated is of definite guarantee. As such, a mobile AR interface to visualise additive manufacturing is proposed. In this work, cloud database is integrated to the AR system to enable online information retrieval on the product material. Overall, a prototype system has been developed and evaluation of the usability of the AR interface has been conducted. The mobile application is essential to guide operators and could possibly be used to design additive manufacturing process.

Keywords Augmented reality · Additive manufacturing · Mobile visualisation · Assisted maintenance · Virtual simulation

1 Introduction

Additive manufacturing (AM) is a process to fabricate parts or products from a 3 dimensional (3D) digital model by depositing the material in layer by layer custom. With the aid of computer-aided-design (CAD) technology, design of a complex 3D structure can then be transformed into a digitalized data for computerized printing

Y. B. Chu (✉) · C. W. Chang
Tunku Abdul Rahman University College, Kuala Lumpur, Malaysia
e-mail: allan.chu.y.b@gmail.com

© The Author(s), under exclusive license to Springer Nature Singapore Pte Ltd. 2021

Z. Zakaria and S. S. Emamian, *Advances in Electrical and Electronic Engineering and Computer Science*, Lecture Notes in Electrical Engineering 741,
https://doi.org/10.1007/978-981-33-6490-5_1

[1]. Manufacturing that incorporates AM aims to minimize the cost of production results from customized fabrication of tailored-made products. Progressively, the technique allows fast printing of complex structures with lower material wastage and subsequently improves the yield of production [2]. Till date, AM has gained much popularity since its first inception in the 1980s. Nowadays, the technology can be found in application such as medical [3, 4], electronics [5], mechanical assembly [6], automotive industry [7] and so on.

Despite the appealing image of AM in the era of industry 4.0, the market is consistently challenged by the continuous changes of customer demand [8]. Generally, these changes are unpredictable and varied through time. Although characterization can be performed to model the specification of customer, mapping the data to the specific production batch for customized fabrication can be very complex. Moreover, in addition to the competitive and fast-growing market, manufacturers are in short to coordinate their manufacturing platform to match these changes in real time. Hence, there is a need to standardize the manufacturing process to be responsive to changes in customer demand and ultimately ensuring customer satisfaction.

Augmented reality (AR) is an emerging technology that creates visual perception of physical object in real-world environment via computer techniques. Particularly in manufacturing, AR helps improve customer satisfaction, creates a much interactive user interface, assist in understanding complex manufacturing operations and also improve efficiency of production plant [9–12]. Recently, researchers have demonstrated a working AR to output context aware information to support maintenance and assembly process [10, 13]. The proposed AR interface is able to improve efficiency of the manufacturing process by means of adaptive knowledge transfer to the worker in real time. Equally, if the information is readily accessible and visualised to customer, uncertainties to the desired specification of the finished product can be managed prior to actual fabrication. This enables the fabrication of customized product in batch production settings and in addition reduces the expenses needed for a similar fabrication in a customized manufacturing plant. On the other hand, with the augmented visualisation of product, customer can then customize the process of additive manufacturing based on own preference and thus improve customer satisfaction.

In view of this, development of an augmented interface to visualise AM process of personalized product is both strategic and advantageous. Therefore, the objective of this research is to design and create a user-oriented augmented display to interface AM for fabrication. Specifically, firebase, a mobile development platform is integrated to the proposed AR for remote data retrieval of AM. In this paper, an implementation of cloud database integrated AR to personalize product is presented. At the same instance, the virtual reality system serves as a visual interpretation, simulation tool as well as modeling platform to characterize 3D prototyping of customized product.

The remainder of this paper is structured as follows: Sect. 2 outlines the related works; Sect. 3 describes the initial implementation of the augmented reality system on Ender 3 3D printer; Sect. 4 demonstrates an evaluation of the augmented reality system in a real-world application; Sect. 5 draws the conclusions, limitations and future improvement of the work.

2 Related Works

The term "augmented reality" was first coined by Caudell and Mizell when their developed an experimental AR system on wiring harnesses [14]. Throughout the decade, it can be observed that AR has been increasingly used for researches in manufacturing industry. In research, the group of A. Y. C. Nee and S. K. Ong has been actively committed to the development of AR system. In 2012, Zhu et al. [15] have developed an authorable context-aware AR system to aid maintenance operator. Using marker based tracking method, the AR content is generated on mobile user interface in real time. Overall, the proposed system enabled the operators to update the rendered AR content on-site and successfully implemented in a CNC milling machine a year later [16]. During the same year, Wang et al. [17] from this group have introduced an interactive AR interface to aid manual assembly of knuckle joint. To augment the assembly sequence accordingly, the group adjust the orientation of the virtual joint based on the orientation of hand movement. This omits the necessity of an auxiliary computer-aided design (CAD) information and reduce the preparation time upon actual assembly. After another year, Fang et al. [18] have presented an AR interface to plan robot path. An Euclidean distance based method has been developed to interface human command to virtual robot in AR environment. Till recently, Wang et al. [9] demonstrated an ubiquitous AR for shop floor planning. In this work, genetic algorithm has been used for dynamic task scheduling of shop floor in real time. The proposed AR system will assign maintenance task to on-site operator should the machine experience breakdown.

Aside that, AR solution has been explored to customize design for children footwear [19]. In this work, marked stocking is used for augmenting virtual footwear on the foot of customer. The environment is consistently captured by a KINECT sensor and as such enabled real time foot tracking of customer. Never the less, maintenance is still the major focus for most AR researches [20]. It is found that the augmented content helps reduce the maintenance time and hence leading to an improved efficiency. Interestingly, some researches claimed that by projecting the augmented content on large screen, the augmented instructions managed to help reduce the completion time of maintaining a motorbike engine [21]. Furthermore, some researchers even converted the maintenance instruction as standard symbol for visualisation [22]. Overall, the researches on AR are mainly augmented on a marked object and are focused on overlapping the information on a real time recorded environment background.

This paper focuses on development of an AR system to visualise additive manufacturing. As mobile devices are used in this work, sophisticated setup such as high end camera and headset is omitted. Moreover, since edge detection is employed to track the target object, physical markers that require precise settings can be excluded as well. This work presents an interactive mobile interface featuring two modes of visualisation: first mode is to augment maintenance instructions onto the virtual manufacturing station. Second mode observes a virtual simulation scene for the additive manufacturing process. The development of AR system in this work is performed using Unity 3D and Vuforia for their robust tracking capability. In addition, Firebase is integrated to the AR to augment specific content from the server onto the mobile screen. For simplicity, Ender 3 3D printer is setup as the additive manufacturing station for this work.

3 Results

The work proposes development and implementation of an augmented interface on mobile devices. The augmented virtual model should visualise maintenance, simulation of additive manufacturing process and contain the integration of cloud server for information retrieval. In this work, One Plus 5 was used to run the developed mobile application. The architecture of the proposed system is illustrated in Fig. 1.

3.1 Target Tracking

The first part of the work is focused on tracking or recognition of target object to be displayed onto the mobile screen. In view of the aforementioned advantages, this paper demonstrated edge detection method instead of the marker based solution.

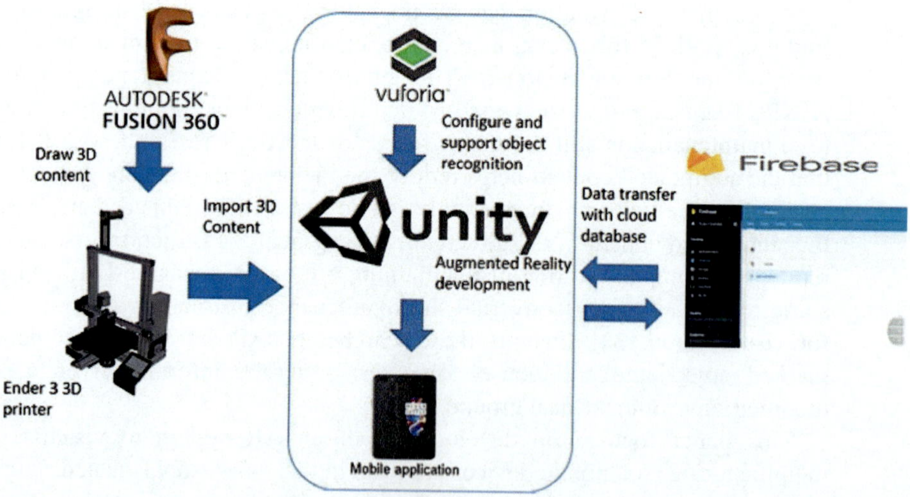

Fig. 1 Architecture of the proposed augmented reality interface

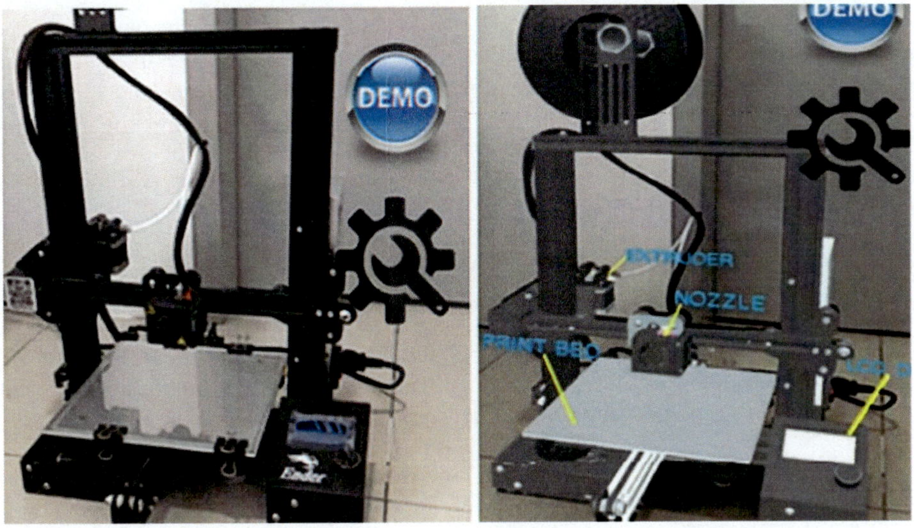

Fig. 2 The AR interface prior to object tracking (left) and after target recognition (right)

Figure 2 (left) presents the AR scene prior to target tracking. Once the edges set in the AR system matches physical outline of the machine, virtual model of the setup will overlap the actual station in real time. To create a much realistic AR scene, the edge layout is rendered invisible on the AR scene [23].

Secondly, the work focused on creating the virtual object of the manufacturing station. To project the virtual model of the target station, the 3D digitalized content of the machine is required. In this work, Autodesk Fusion 360 was used to design and construct the 3D content of the manufacturing station. Based on Fig. 1, Vuforia engine is integrated to enable object recognition and robust tracking onto the Unity 3D development platform. The virtual model shown in Fig. 2 (right) is the 3D content drawn using Fusion 360.

3.2 Assisted Maintenance

This section visualises the maintenance and servicing of the 3D printer. Generally, the maintenance information is augmented as text on the AR scene to suggest solution for the particular maintenance activities. The instructions provided on the maintenance activities are based on the possible causes of failure and its corresponding resolve action. In cases of breakdown, the activities shown on the augmented instructions should be conducted. Instead of manual diagnostic, the AR system illustrated graphical presentation of the augmented command for assisted maintenance. Since the time for lookup of manual is reduced, the downtime of maintenance is also decreases and hence leading to higher efficiency of additive manufacturing. To improve the interaction with user, the augmented content is presented with symbol, textual and intuitive presentation. While symbol and textual describes the maintenance information as augmented logo and instruction respectively, intuitive presentation allows better understanding of the maintenance instructions with projected graphical content.

4 System Implementation

The intuitive presentation of the maintenance instruction is realized with object transformation of the manufacturing station. To convert manufacturing station from position 1 to position 2, an object transformation matrix is applied.

$$\boldsymbol{O} = \boldsymbol{T} * \boldsymbol{o} \quad (1)$$

$\boldsymbol{O}$ is the new coordinate of object after oriented
$\boldsymbol{T}$ is the transformation matrix
$\boldsymbol{o}$ is the original position or orientation of object in world coordinate.

The transformation matrix $\boldsymbol{T}$ consists of a 3 × 3 rotational matrix r and translational vector t.

$$\boldsymbol{O} = \begin{bmatrix} \boldsymbol{r}_{11} & \boldsymbol{r}_{12} & \boldsymbol{r}_{13} & \boldsymbol{t}_x \\ \boldsymbol{r}_{21} & \boldsymbol{r}_{22} & \boldsymbol{r}_{23} & \boldsymbol{t}_y \\ \boldsymbol{r}_{31} & \boldsymbol{r}_{32} & \boldsymbol{r}_{33} & \boldsymbol{t}_z \end{bmatrix} * \begin{bmatrix} \boldsymbol{X} \\ \boldsymbol{Y} \\ \boldsymbol{Z} \\ 1 \end{bmatrix} \quad (2)$$

The system calculates the destination coordinate of the target object based on the transformation matrix. The outcome of this stage observes the object oriented in the AR scene. This approach omits complicated mapping of marker frame and the need for calibration of high-tech camera.

4.1 Maintenance of Nozzle

The maintenance operation of the nozzle is consisting of 3 steps. Figure 3 shows the maintenance scene of the proposed AR system. The operation involved unlocking the casing and fan cover to reach the nozzle. Graphical content of the parts and servicing steps were projected onto the AR scene. Servicing of the nozzle is essential since the filament for printing often clog at the nozzle and causes breakdown of the machine. In order to validate the developed AR mobile interface, a group of engineers aged within 22–29 years old were requested to perform the maintenance task using the AR system. A survey which scaled from 1 (disagree) to 10 (agree) was given to the engineers and none of them had any experience on using AR for visualisation. Since manual lookup is omitted with the application of AR, the time on performing the task is expected to be shorten and hence increasing the efficiency of the maintenance task. Figure 4 shows the response of participants on the effectiveness of the AR system in improving efficiency of the maintenance operation.

Based on the response, it is worth noticing that all of the participants agreed on using the proposed AR system in improving the efficiency of maintenance. It was also found that the participants are more confident on servicing the nozzle using the AR system. In view of this, the developed AR interface is proved to be useful especially in training new staff on the aforementioned task.

4.2 Virtual Simulation

This section presents a virtual scene for user to visualise the additive manufacturing process. Figure 5 presents the virtual simulation scene for the additive manufacturing process. The AR scene allows user to visualise the process from various angles by dragging the screen in accordance to the finger movement. For simplicity,

Fig. 3 Augmented reality presentation of the maintenance task

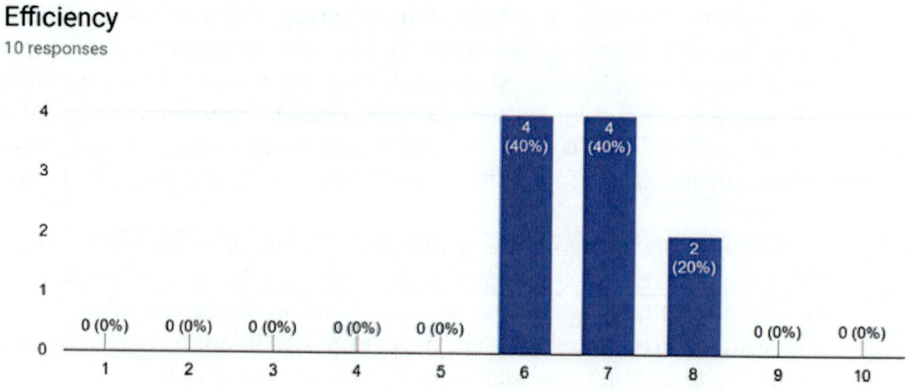

Fig. 4 Response of participants on the effectiveness of the developed AR system

this section simulates the product in shape of a 1000 cm^3 sized cubic. The rate of filament insertion in this simulation was set at 6.67 filament/sec and the filament size as well as printing height is of 0.2 mm^3.

The virtual model can easily be dragged into the station for simulation. However, due to the technical setting of the development platform, the 3D content

Fig. 5 Virtual simulation scene of the additive manufacturing process

cannot be automatically loaded into the virtual scene, and hence the virtual model to be visualised on simulation should be integrated into the platform in advanced.

Some technical difficulties were faced during the simulation. Foremost, the AR system experiences lagging upon simulation of the virtual models onto the scene which may be caused by the low memory or processing power of the mobile device used. Thus, for smoother operation, only one model (blue cube) is simulated on the AR scene. This section is important as it provides visualisation on the additive manufacturing pathway and the number of steps required to print the product. The information is especially useful to plan and design the manufacturing process with respect to the product requirement. A similar survey which is scaled from 1 (agree) to 10 (disagree) was given to the participants to assess their experience in using the AR interface. From the survey, it can be evaluated that the mobile AR interface is easily used and comfortable for usage to user. The participants also responded that, the AR system assisted them to properly visualise the technical aspect of the fabrication and the desired product feature prior to actual fabrication. Figures 6 and 7 shows the response of participants on the feature and conveniency of the developed AR system respectively. As the AR provides virtual simulation, misunderstanding of product enquiry and wrong fabrication of unwanted product can be prevented. The AR interface is compatible for installation on mobile device and convenient for visualisation at any time. Never the less, further standardization of the AR interface and simulation on sophisticated product is still required in the future.

4.3 Cloud Database

In this work, the AR interface is integrated with a cloud database for online information retrieval. Figure 8 presents the context ontology for the cloud database service.

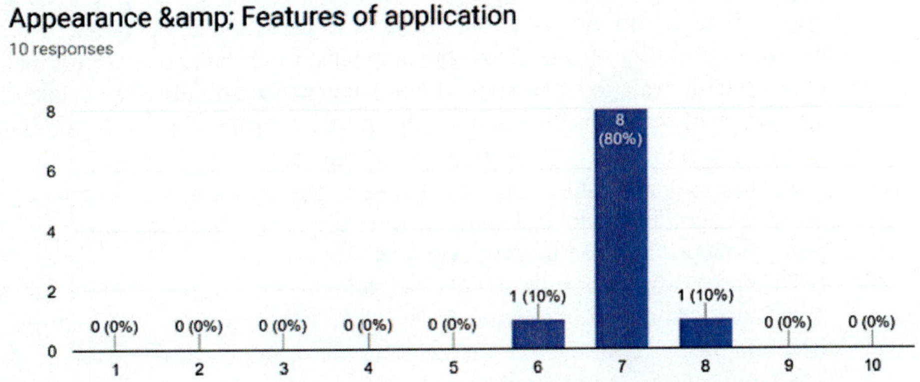

Fig. 6 Response of participants on the appearance and feature of the developed AR system

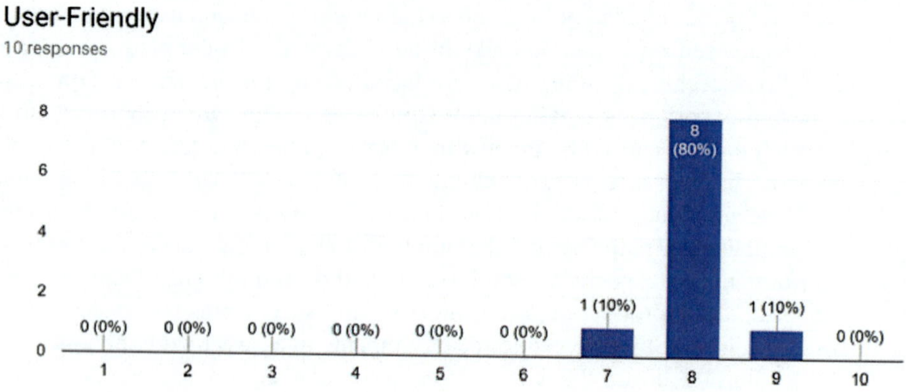

Fig. 7 Response of participants on the conveniency of the developed AR system

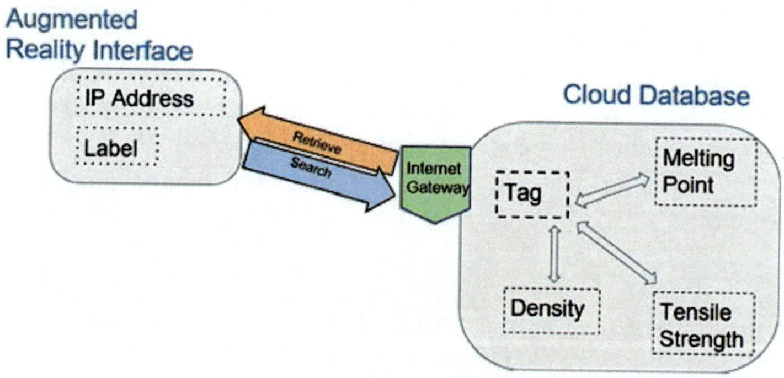

Fig. 8 Context ontology of the cloud database service

An internet connection is required to access the content in database. For better user experience, the internet address to retrieve the information has been preliminary hardcoded into the AR system. User is required to key in the name of the product material on the search bar in order to activate this operation. The label in the AR scene should match the tag referring to the respective information and output the information onto the mobile screen accordingly. For implementation purpose, only basic information of the product material such as melting point, tensile strength and density were programmed into the mobile application. Once a valid entrance has been keyed, the developed application will automatically retrieve the information from cloud. Figure 9 shows the coding used to retrieve information from the cloud database to the developed AR system.

The above work presents an implementation of a markerless mobile AR interface to visualise additive manufacturing. There are two important contributions; (1) assisted maintenance on the 3D printer using the developed AR interface, (2) virtual simulation of the additive manufacturing process. The developed AR allows retrieval of technical information related to the manufacturing process from a

```
private void GetDataFromFirebase()
{ RestClient.Get<filamentInfo>(url: "https://unitytest2bbdd.firebaseio.com/" +
        searchFilament.text + ".json").Then(onResolved: response =>
    {
        user = response;
        UpdateScore();
    }); }
```

Fig. 9 Code to automatically retrieve information from the cloud database

dedicated cloud database. The integration is meant for implementing Industry 4.0 into the AR technology. Never the less, further development and research is still required in order to develop and standardize the use of AR in the manufacturing industry. The developed AR is able to visualise the additive manufacturing process of desired product on own try-on. Overall, credits to the mobile framework, the developed mobile AR interface omit the use of expensive camera and sophisticated maker alignment. Based on the survey, the mobile application developed is in agreement useful in terms of improving efficiency and in guiding staff especially to first-time learner. On the whole, the proposed AR system has been developed and the functions have been tested accordingly.

5 Conclusions

In the presented work, AR was implemented to visualise additive manufacturing process. The proposed approach incorporated virtual simulation for user to visualise the production. In addition, database was integrated into the AR for data transfer purposes. As there is a growing interest of manufacturing trend shifting to product customization, the developed AR system may serve as solution to interpret customer demand into product specifications. Based on the evaluation, although AR is not dominantly used in manufacturing, it may prove helpful in the field of design for additive manufacturing. Especially in the consultation stage, the use of AR can help avoid misunderstanding in the product requirement which may lead to additional costs for reproducing the product.

Future work will focus on expanding the feature of AR. Expansion to include new materials such as biomaterials, metals and alloys and their specification in additive manufacturing is anticipated. The development of AR on additive manufacturing of highly customized product for various functions and applications is also part of the research interest. By looking at the system performance, research on hardware which supports the AR is also important. The research on testing and application of AR in additive manufacturing still require further standardization.

References

1. Chergui A, Hadj-Hamou K, Vignat F (2018) Production scheduling and nesting in additive manufacturing. Comput Ind Eng 126:292–301
2. Culmone C, Smit G, Breedveld P (2019) Additive manufacturing of medical instruments: a state-of-the-art review. Addit Manuf 27:461–473
3. Javaid M, Haleem A (2018) Additive manufacturing applications in medical cases: a literature based review. Alexandria J Med 54(4):411–422
4. Javaid M, Haleem A (2019) Current status and challenges of additive manufacturing in orthopaedics: an overview. JCOT 10(2):380–386
5. Tan HW, Tran T, Chua CK (2016) A review of printed passive electronic components through fully additive manufacturing methods. Virtual Phys Prototyping 11(4):271–288
6. Sossou G, Demoly F, Montavon G, Gomes S (2018) An additive manufacturing oriented design approach to mechanical assemblies. J Comput Des Eng 5(1):3–18
7. Leal R, Barreiros FM, Alves L, Romeiro F, Vasco JC, Santos M, Marto C (2017) Additive manufacturing tooling for the automotive industry. Int J Adv Manuf Tech 92:1671–1676
8. Roblek V, Mesko M, Krapez A (2016) A complex view of industry 4.0. SAGE Open 6(2):1–11
9. Wang X, Yew AWW, Ong SK, Nee AYC (2020) Enhancing smart shop floor management with ubiquitous augmented reality. Int J Prod Res 58(8):2352–2367
10. Erkoyuncu JA, Amo IFD, Mura MD, Roy R, Dini G (2017) Improving efficiency of industrial maintenance with context aware adaptive authoring in augmented reality. CIRP Ann Manuf Techn 66(1):465–468
11. Flavián C, Ibáñez-Sánchez S, Orús C (2019) The impact of virtual, augmented and mixed reality technologies on the customer experience. J Bus Res 100:547–560
12. Syberfeldt A, Danielsson O, Holm M, Wang L (2015) Visual assembling guidance using augmented reality. Procedia Manuf 1:98–109
13. Osorio-Gomez G, Vigano R, Arbelaez JC (2016) An augmented reality tool to validate the assembly sequence of a discrete product. IJCAET 8(1–2):164–178
14. Bottani E, Vignali G (2019) Augmented reality technology in the manufacturing industry: a review of the last decade. IISE Trans 51(3):284–310
15. Zhu J, Ong SK, Nee AYC (2013) An authorable context-aware augmented reality system to assist the maintenance technicians. Int J Adv Manuf Tec 66:1699–1714
16. Zhu J, Ong SK, Nee AYC (2015) A context-aware augmented reality assisted maintenance system. Int J Comput Integ M 28(2):213–225
17. Wang ZB, Ong SK, Nee AYC (2013) Augmented reality aided interactive manual assembly design. Int J Adv Manuf Tech 69:1311–1321
18. Fang HC, Ong SK, Nee AYC (2014) A novel augmented reality-based interface for robot path planning. Int J Interact Des Manuf 8:33–42
19. Luh YP, Wang JB, Chang JW, Chang SY, Chu CH (2013) Augmented reality-based design customization of footwear for children. J Intell Manuf 24:905–917
20. Mourtzis D, Zogopoulos V, Vlachou E (2017) Augmented reality application to support remote maintenance as a service in the robotics industry. Procedia CIRP 63:46–51
21. Fiorentino M, Uva AE, Gattullo M, Debernardis S, Monno G (2014) Augmented reality on large screen for interactive maintenance instructions. Comput Ind 65(2):270–278
22. Scurati GW, Gattullo M, Fiorentino M, Ferrise F, Bordegoni M, Uva AE (2018) Converting maintenance actions into standard symbols for augmented reality applications in industry 4.0. Comput Ind 98:68–79
23. Brito PQ, Stoyanova J (2018) Marker versus markerless augmented reality. which has more impact on users? Int J Human-Comput Int 34(9):819–833

Safety Risk Assessment and Warning System at Engineering Laboratory Through IoT Technologies

Seerla Kanagarajoo Betharajoo, Sivajothi Paramasivam, and H. S. Chua

Abstract Common contributing factors associated to accidents or near-misses in laboratories arises due to risks related to improper use or maintenance of laboratory test equipment and machine. Hence, maintenance and regular inspection of laboratory test equipment and machine are essential parts of safety activity. The project exploits dominance of Internet of Things (IoT) used in many applications to leverage combination of networks and radio frequency identification (RFID) tags technologies to gauge safety risks associated with laboratory inventory in an automated manner. Thus, this paper makes the following contributions:(1) developed an automated inventory laboratory management system that enable test and measurement equipment and machine to be traceable by RFID technology (2) generated an analysis output using informatics as a diagnostic component to improve safety risks presented by test and measurement equipment and machine and (3) developed a graphical user interface that served as the graphical output in monitoring and controlling safety status. Critical issues associated with safety risks which include maintenance schedules, calibration dates, outdated equipment and inventory of equipment are gauged by this system. Subsequently it successfully provides graphical notification on each issue which reflects intact, caution and warning status with green, yellow and red colour codes respectively. In addition, graphical information highlights percentage of progress to stress the severity of the evaluated issue to the technical personnel. This smart system initiates electronic-mail notification to prompt proactive actions to be taken by the technical personnel in a timely manner to mitigate potential accidents or near-misses within the laboratory settings.

Keywords Inventory management · Engineering laboratory · Internet of things · RFID · Networks · Safety risk

---

S. K. Betharajoo · S. Paramasivam (✉) · H. S. Chua
School of Engineering, University of Wollongong Malaysia KDU, Glenmarie Campus, Shah Alam, Selangor, Malaysia
e-mail: siva@kdu.edu.my

© The Author(s), under exclusive license to Springer Nature Singapore Pte Ltd. 2021
Z. Zakaria and S. S. Emamian, *Advances in Electrical and Electronic Engineering and Computer Science*, Lecture Notes in Electrical Engineering 741,
https://doi.org/10.1007/978-981-33-6490-5_2

1 Introduction

As teaching and research activities are becoming more complex; academic institutions are aware of and acknowledge that experiments have not always been done without laboratories accidents and near-misses. The reason being problems with test equipment and machine may present in many ways during laboratories session which potentially pose safety hazard such as danger of electrical shock, and some equipment can cause physical cuts and emit dangerous sparks. Common contributing factors associated to laboratories accidents or even near-misses arises due to risks related to improper use or maintenance of laboratory test equipment and machine. Hence, maintenance and regular inspection of laboratory test equipment and machine are essential parts of safety activity.

As technology has advanced, test and measurement equipment are developed for measuring ever more complex parameters to ever increasing levels of precision and accuracy. Thus, these come at an increased cost and large investment for both acquisition and maintenance. The situation is further compounded as majority of acquired test and measurement equipment of advanced technologies and developing materials introduce new safety and quality risks that has high potential to create harm to stakeholders involving academic staffs, technicians, students, apprentices, administrative staffs, short term visitors as well as vendors and suppliers if not properly used or maintained [1].

Studies have indicated that in many engineering teaching and research laboratory settings safety culture has been the prime focus to prevent accidents or injuries as these occur through lack of knowledge of safety risks involving test and measurement equipment and machine. According to a student was killed by a machine–shop equipment while carrying out experiments in the laboratory [2, 3]. An alarming finding revealed that laboratory accidents, both in teaching and research laboratories, represented 18.4% of the total incidents reported at Iowa State University from 2001 to 2014, and dominantly victims involve students and employees [4]. Consequently, gravely emphasizes that academic laboratories are more dangerous because of the lacklustre attitude approach towards safety specifically in maintaining functional quality of various test and measurement equipment and machine in an effective and systematic manner.

Laboratory management is a very important element specifically in ensuring safety aspect of academic and research laboratory activities. Most cases in the past event where accidents in laboratory happened are due to the failure of keeping track those faulty and non-functional test and measurement equipment and machine, which are also closely related to the poor inventory and maintenance system.

It is against this backdrop that this paper makes the following contributions: (1) developed an automated inventory laboratory management system that enable test and measurement equipment and machine to be traceable by RFID technology (2) generated an analysis output using informatics as a diagnostic component to improve safety risks presented by test and measurement equipment and machine and (3) developed a graphical user interface that served as the graphical output in monitoring and controlling safety status. Thus, it is envisaged that efficient inventory maintenance and control implementation would yield a high degree of accountability in detecting risks and safety related to use of test and measurement equipment and machine.

2 Academic Laboratories—Significance on Assessment of Resources

Laboratory is as an integral part of academic resource in engineering education as it forms as the primary vehicle where basic experimental skills are learnt by systematically performing a set of prescribed and suitably designed experiments. As such, these laboratories supporting various engineering disciplines houses an array of hand tools, test and measurement equipment and machine.

In these settings, students are to carry out experiments by making all the required connections of the laboratory equipment, and then they manually operate tools and machinery in order to execute experimental exercises. Thus, prompts a query whether usage of various test and measurement equipment and machine meets the desired safety compliance in preventing accident or severe injuries during laboratories session [1].

In this connection, safety risks associated with array of hand tools, test and measurement equipment and machine being used have received the most attention in the laboratory safety. As shown in Table 1, the most common equipment-related hazards in laboratories come from devices powered by electricity devices for work, materials driven by compressed gases or chemical and rotating equipment and machines or tools for cutting and drilling which account for the greatest frequency of laboratory accidents and injuries [5].

In light of the severity of the given cases, risk assessment on safety hazards within academic and research laboratory settings are the responsibility of all stakeholders specifically the technical support team who are tasked to ensure tools, test equipment and machines are always in healthy working conditions [9]. Consequently, safety risk can be identified in the event laboratory personal may

Table 1 A partial list of laboratory accidents at academic institutions (2006–2019) [6–8]

Year	Institution	Accident description
2006	Cleveland State University, USA	Staff died of electrocution of electrical device
2007	Indian Institute of Science, India	Researcher died of laboratory explosion
2009	Army Base—Iraq	Staff died of electrocution by machine
2011	Yale University, USA	Undergraduate died by rotating axis of lathe machine
2019	Technion, Israel Institute of Technology	Staff died due to explosion during materials experiments

notice or notified subtle changes such as drift in quality control or calibrator values, or obvious flaws in equipment function. This also entails soliciting feedback from co-workers and colleagues to improve safety and the process as there will be many errors and problems occurring that may go undetected. Thus, proper maintenance work of test and measurement equipment and machine in the laboratory is necessary to ensure accurate and reliable usage for experimental activity and learning needs with acceptable degree of safety aspect.

Thus, these arrays of resources require efficient and aptly smart system with a provision of monitoring and early alerting mechanism of flaws in equipment function or related issues that has direct bearing on addressing laboratories accidents and near-misses. As such, it is imperative that these academic resources are well maintained, serviced, tracked and recorded accurately in proper time schedule [10].

Engineering and research–related laboratory environment may involve a wide variety of safety hazards which include unguarded machinery and moving machinery parts; electrical hazards from test and measurement equipment and machine as well as malfunction and outdated equipment. In this relation, realization of development of safety control measures involving test and measurement equipment and machine in the inventory systems should be maintained and serviced periodically to ensure that laboratory services consistently meet the needs of teaching and learning situations at all times.

Conversely, the analysis of a well-defined scope of work in evaluating element of risk assessment that has direct bearing on safety hazard issues involve:

- maintaining good-quality tools, devices, equipment and machine
- removing and disposing of outdated equipment from use
- schedule calibration and performance evaluation
- perform maintenance work of equipment
- identifying equipment for repair work due to breakdowns and failures
- sustaining quality compliance on record keeping.

Well-designed laboratory facility that houses all the test and measurement equipment and machine adequately alone is insufficient to address improvement effort on safety risk to minimize or eliminate exposures of accident or injuries to laboratory personnel, community and environment. In fact, the laboratory management system, in its quest to formulate improvement effort on safety assessments and mitigation plans; documents and records of laboratory set ups are essential part of the inventory system.

Consequently, in order to retain and develop safe work performance within the engineering laboratory settings; use of informatics forms as a highly desirable solution to evaluate the risks presented by safety hazards.

In this relation, leveraging on development of Internet of things, quality assessment of inventory tracking and maintenance monitoring and early alerting mechanism can be automated by computerizing using cloud architecture and radio frequency identification (RFID) technology. The reason being the advanced technology of RFID provides an efficient mode of assessing the risk revolving various equipment as it manage and communicate information on inventory details and specifications. In fact, RFID has gain increased attention recently for its use on remote storage and retrieval of data for many applications such as warehouses, production shop floors, logistics companies, distribution centers and retailers [11].

In addition, the graphical representation provides unique colour code indicator which uses colour codes to reflect status of the test and measurement equipment based on the level of risk and safety. In this relation, red colour is interpreted that it is in a critical level which will indicate on the service or is beyond repair condition. The yellow colour signifying a warning sign that the equipment service is almost due for service and maintenance. As of green, it is safe to operate the test and measurement equipment.

Thus, this entails embracing smart technology to automate and rely on an information system in critical decision making which is viewed as an effort on modernizing the quality system which has direct bearing on safety aspect of stakeholders.

3 Methodology

Although the laboratory adheres to basic safety regulations, this study is intended to further enhance improvement on safety risk of the test and measurement equipment through usage of informatics and RFID technology [12]. On the informatics section, visual basic is used to design and develop the GUI interface to relay vital information related to status of the test and measurement equipment via network. There will be two prong of monitoring whereby the first is to monitor on the safety risk status related to each test and measurement equipment in terms of maintenance and usage and the second is on the inventory management in meeting quality compliance on record keeping.

Basically, it keeps track of the equipment due date in terms of calibration date, maintenance date, retirement date and the list of equipment in the inventory. It alerts the lab support team if the date is approaching closer to due date.

3.1 Project Design and Settings

The layout design and installation of RFID system was tested in Electrical and Electronics Systems Laboratory which is located at the second floor of the University Campus. This particular laboratory houses many test and measurement equipment of different brands as well as technical specification. The complete floor layout of the laboratories under the School of Engineering is shown in Fig. 1.

The RFID antenna is mounted beside the exits of the laboratory whereby a pair of brackets are used to hold the antenna at an angle of 45° position to capture the

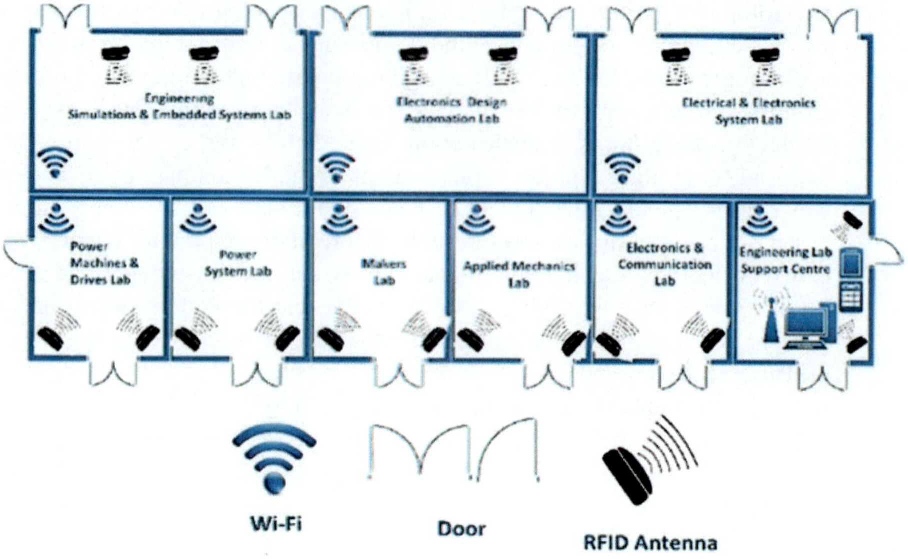

Fig. 1 Floor plan of all engineering laboratory system

best frequency range for the tags on the equipment. This particular laboratory houses 36 different test and measurement equipment which are tagged with a unique identification. The frequency range for this set up falls within 865–867 MHz frequency. A proposed distance between the two brackets are 2 m range and the height is in 2.5 m. A single reader is used and operate on a wireless mode to capture the data from the RFID antenna. This will well cover the area of the workstation containing the said number of equipment [13].

In the operational mode, the RFID reader performs a real time scan to check the availability of the test and measurement equipment at its designated workstation. In fact, a routine check on a timely manner is done to ensure their presence for inventory management. This is done by tracking the RFID tag which consist a unique identity for each equipment. This is shown in Fig. 2.

Fig. 2 The actual view of RFID antenna in electrical and electronic laboratory

Dominantly, the process involves, detection of a passive tag where by information in the tag will be read and send to RFID reader. In turn, it will be transmitted to the wireless link communication port. The data is transmitted via wireless router to a client computer in the engineering lab support centre which is located about four meters away. The data will updated in the client computer which is supported by a graphical user interface. The system configuration enables the users to access into the informatics system to do data retrieval or modification of data for analysis. The flow chart of this mechanism is shown in Fig. 3.

This entails the system to generate an analysis output using informatics as a diagnostic component by checking critical information pertaining to safety risk of various test and measurement equipment at the workstation. In fact, all this are able to be monitored using network and subsequently prompts vision base notifications onto the client computer of the laboratory support team.

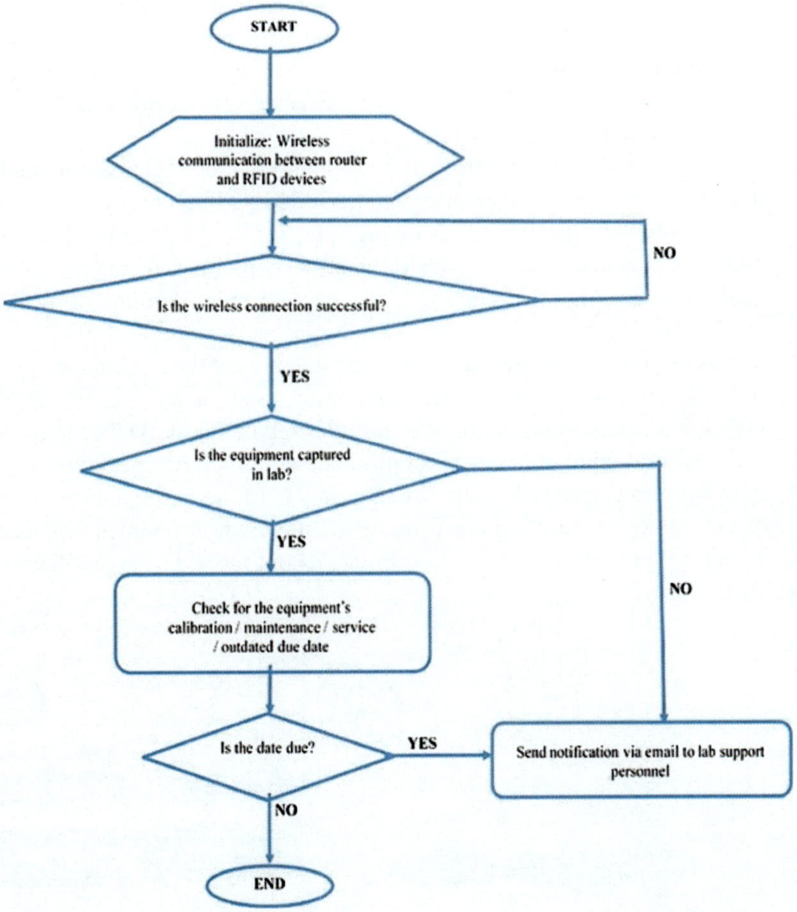

Fig. 3 The flowchart of the system

In addition the system updates the inventory database and eventually triggers a warning alert as part of security measure to notify if in the event of the equipment is absent at the workstation. As indicated, the system design uses a database to exhibit graphical output of safety risk status of test and measurement equipment that is not functioning and is beyond repair status, or when it is outdated lifespan and should be replaced and to the extent of calibration due of each equipment that are in the calibration list of the database.

3.2 *Database Design Settings*

The database system authorized to act both as an administrator and the user rights. In administrator mode, it permits to do all function including add or remove users and to update the inventory status, which includes borrowing out of the laboratory tools. In the user mode, it only permits to check the database, borrowed out of tools and receive the tools after usage of experimental session. In ensuring effective system control, only the administrator and assigned users have access rights.

Figure 4 shows the graphical user interface of logging in and main menu of the created the database system design.

As shown in Fig. 5, there will be four sub menus. Each sub menu has a database that performs the coded relating to its category. In fact, acts as a platform to view various relevant details which include the stock check to the borrowed details by authorized stakeholders. Critically, it also keeps track of maintenance, service and calibration due dates of each test and measurement equipment to the extent of the retiring due date.

In this graphical representation, critical details of the equipment's safety risk status is verified and assessed to reflect the condition of the equipment such that when it is faulty, is it a beyond repair status or under repair status. This is shown in Fig. 5.

In calibration alert page, calibration due dates will be checked and prompts timely indication to the relevant superior. Similarly, in the maintenance and service menu, dates would be checked on routine maintenance schedules and initiates alerting signal on a timely manner. This applies to the retiring (beyond lifespan) report.

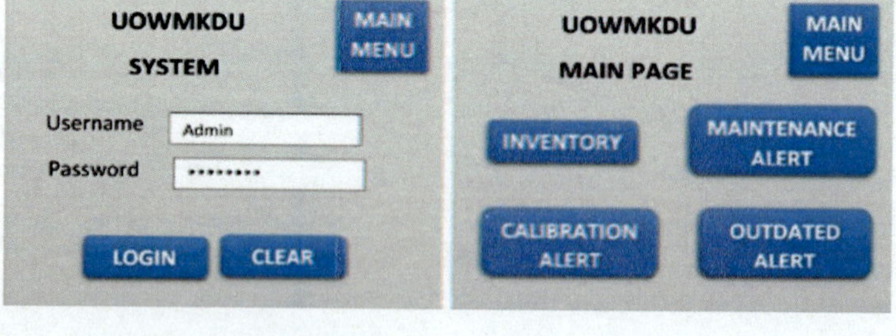

Fig. 4 Login page and main page

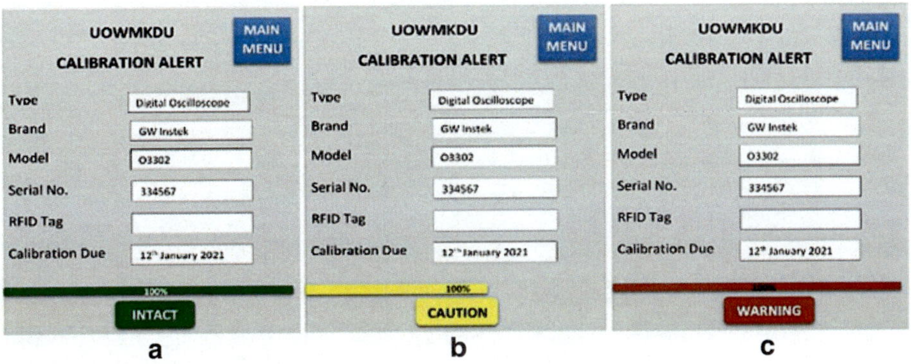

Fig. 5 Calibration alert page

In this setting, when the current date is within the due date, a green colour indicator is produced with a message "INTACT" is shown as illustrated in Fig. 5(a). In the event, the current date is approaching nearer to the due date, it will highlight yellow colour with a message "CAUTION" as illustrated in Fig. 5(b). Finally, when it is actually the due date or passed the due date, it will highlight red colour and shows "WARNING" message as illustrated in Fig. 5(c).

In addition, there will be a progress bar which will indicate of the time of expiry/due for the identified test and measurement equipment.

Under the maintenance alert page, maintenance and service schedules are crossed checked to ascertain and prompt these activities accordingly, Similarly, when the date sufficiently from due date, it will highlight green and display "INTACT" message as illustrated in Fig. 6(a). In the second scenario, when the date is approaching the due date, it will highlight yellow and show the message "CAUTION" as illustrated in Fig. 6(b) and when it is due date or pass due date, it will highlight red as and show the message "WARNING" illustrated in Fig. 6(c).

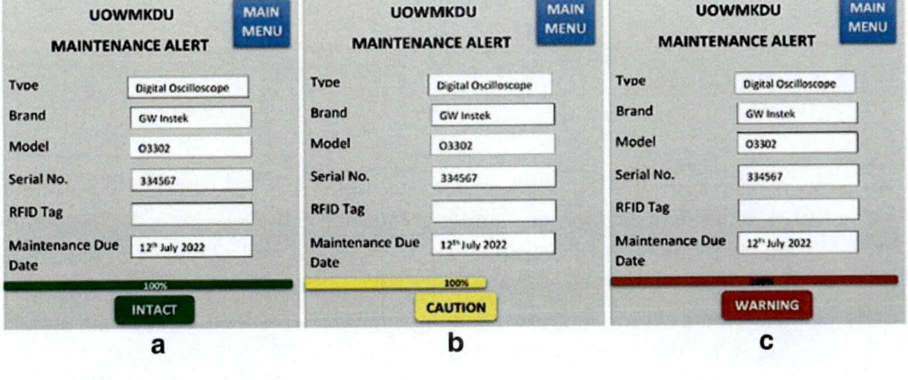

Fig. 6 Maintenance alert page

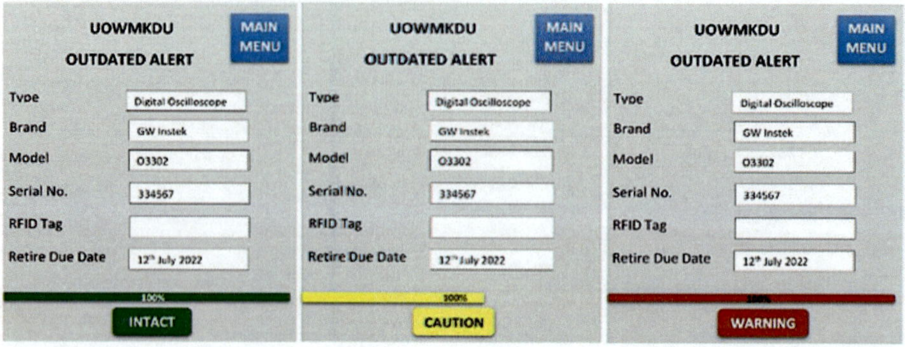

Fig. 7 Outdated alert page

In addition, there will be a progress bar which will indicate of the time of expiry/ due for the identified test and measurement equipment.

In outdated alert page, the system will conduct a check on the purchase date to ascertain the life span of each test and measurement of the equipment. Base of the gathered data, notification will be prompted to highlight date of retirement has been reached and need to be removed from the workstation. As the equipment ages, the system will highlight the colour and message as illustrated in Fig. 7.

If the equipment is within the workable lifespan, the colour will be highlighted in green. In the second condition, if the lifespan is nearing expiry date, yellow colour is produced and over the lifespan reflects red colour. In addition, there will be a progress bar which will indicate of the time of expiry/due for the equipment.

As in any normal database system, it will store all the stock list and keep track of all available items at the lab. Besides that, it will keep track of all loaned items that have been loaned out to all staff and students. This will also show who is the person in charge who had released the items, whereby in such practice, the person will take responsible for any items that has been released out. If the item request is by an academic staff, therefore the responsibility can be set to that person and partly with the support person in charge of releasing the items.

4 Conclusion

- As an overall view, the system monitors critical status of various test and measurement equipment and update the lab support team on status which include due date of calibration, maintenance and service for individual equipment and machine as well as outdate equipment.
- This is to avoid any form of danger to occur as a safety precaution against the usage by the students and staffs.

- Besides that, this is also to monitor the location of the equipment or machine in order not to be misplaced or mixed up in another location.
- This system is developed with element of smart concept which would eventually triggers notification and generate different reports; thus ease the task of technical support team in sustaining quality compliance on record keeping and addressing issues on safety risk efficiently.
- As of the future development of the current system, an integration of smart application would be proposed for the mobile based application to link between the current system with the mobile application [14].

Acknowledgements The authors would like to thank Engineering Support Team, School Of Engineering for their kind support. We would like to acknowledge the equipment and technical knowledge support by University Of Wollongong Malaysia KDU, Glenmarie Campus.

References

1. Annisa N, Puteri F, Nurcahyo R (2018) Safety perceptions in university teaching laboratory
2. Meyer T (2012) How about safety and risk management in research and education? Procedia Eng 42:854–864. https://doi.org/10.1016/j.proeng.2012.07.478
3. Cho Nam, Ji Yong Gu (2016) Analysis of safety management condition & accident type in domestic and foreign laboratory. J Ergon Soc Korea 35:97–109. https://doi.org/10.5143/jesk.2016.35.2.97
4. Menard A Dana, Trant John (2019) A review and critique of academic lab safety research. Nat Chem 12:1–9. https://doi.org/10.1038/s41557-019-0375-x
5. Roy K (2016) Responding to laboratory accidents. Sci Scope 39(9)
6. Foderaro LW (2011) Yale student killed as hair gets caught in lathe. The New York Times
7. Bronstein S (2009) Multiple failures led to Iraq electrocution, pentagon says. CNN.com
8. Miller JD (2006) Ohio cites university in death. The Scientist Magazine®
9. Osang JE (2013) Evaluations of the effect of workshop/laboratory accidents and precautionary steps towards safety practice. IOSR J Electron Commun Eng 6(3):16–22
10. Khalidand M, Kadni T (n.d.) Safety enhancement through calibration and maintenance
11. Mohamad Z (2012) Laboratory quality management requirements of engineering at the polytechnics Ministry of Higher Education Malaysia
12. Hellström D, Wiberg M (2010) Improving inventory accuracy using RFID technology: a case study. Assembly Autom 30(4):345–351. https://doi.org/10.1108/01445151011075807
13. Chen ZN, Qing X (2010) Antennas for RFID applications. In: 2010 International workshop on antenna technology (iWAT). https://doi.org/10.1109/iwat.2010.5464865
14. Park S, Kim J, Kim J (2015) SLAN based user-customized cloud interface sharing for smart mobile devices. In: Kim K, Wattanapongsakorn N (eds) Mobile and wireless technology 2015. Lecture Notes in Electrical Engineering, vol 310. Springer, Berlin, Heidelberg

Optimization of PV Power Output Using Single-Axis Sun Light Tracker and MPPT Algorithm

Ryan Matthew Long, Ahmed. O. MohamedZain, H. S. Chua, J. Jamaludin, and K. M. Yap

Abstract The growth of global energy consumption and the depletion of the non-renewable recourses had caused many environment issues. These environmental issues and the demand of global energy had driven up the implementation of solar energy. The optimization of PV power output can be achieved using solar tracking and MPPT algorithm. Comparison of power output had been made with a system that does not have solar tracking to show the improvements made using a single axis solar tracking system. Power of 100 W solar tracker and two 50 W stationary panels connected in parallel. The values of the resistors of 4 and 8 **Ω** have been chosen based on the solar panels R characteristics. The system consists of three (3) important parts, software hardware, and electronic circuits. DC motor is used to rotate the solar PV panel (100 W) and Two (2) light dependent resistors (LDR) were used as light detectors. Solar tracker and stationary panels facing the East-West direction at University of Wollongon Malaysia KDU, Glenmarie Campus at the coordinates (3.092008, 101.559373). Two prototypes are built for data collection based on Arduino Mega 2560. An increase of 35–40% in power output compared to fixed rooftop installation.

Keywords Solar PV panel · Arduino mega controller · MPPT

1 Introduction

In this current era, there is an increasing need for energy, decrease in supply of fuel and more concerns about environmental issues when using non-renewable energy. This demand is shown in Fig. 1. It projects that the world energy consumption will

R. M. Long · H. S. Chua (✉) · J. Jamaludin
School of Engineering, University of Wollongon Malaysia KDU, Glenmarie Campus, Shah Alam, Malaysia
e-mail: hs.chua@kdu.edu.my

Ahmed.O. MohamedZain · K. M. Yap
HUMAC Research Centre School of Science and Technology, Sunway University, Bandar Sunway, Subang Jaya, Malaysia

© The Author(s), under exclusive license to Springer Nature Singapore Pte Ltd. 2021

Z. Zakaria and S. S. Emamian, *Advances in Electrical and Electronic Engineering and Computer Science*, Lecture Notes in Electrical Engineering 741,
https://doi.org/10.1007/978-981-33-6490-5_3

increase by 28% from 2015 to 2040 [1]. With high cost, limited sources of fossil fuels and worldwide primary energy consumption has grown by 1.8% in 2012 alongside with the need to reduce greenhouse gasses emission, renewable energy resources have become more attractive in the world's energy-based economics [2]. This has led researchers to find an alternative source of energy, which is renewable, and have less negative impacts on the environment such as photovoltaic/thermoelectric generator system (PV-TEG) [3]. In order to reduce the temperature of the PV cell, other researchers has investigated the effect of the temperature on the efficiency of the PV cell by conducting different types of methods [4]. Since the high temperature has a strong effect on the PV cell [5]. There are many renewable sources of energy such as Biomass energy, geothermal energy, Hydropower energy, Marine energy, solar energy and wind energy.

The Fig. 1 shows the ability of renewable energy sources to provide over 3000 times the current global energy needs and solar resource is the highest contributor with an amount of 2850 times the current global energy needs of 2014 [2]. This is because it is abundant across the Earth. The Photovoltaic cells also reduce the amount of carbon dioxide emissions. It is expected that by year 2030, the annual reduction rate of carbon dioxide due to PV cells may be around 1 Gton/year. This is the same amount of India's total carbon dioxide emission in 2004 or carbon dioxide emissions due to 300 coal plants. Many experts state that energy obtained from PV cells will be a very important renewable energy source until the year 2040 [6]. Solar energy is chosen as the energy to be optimized mainly because of its potentially infinite amount of energy supply, causing no air or water pollution.

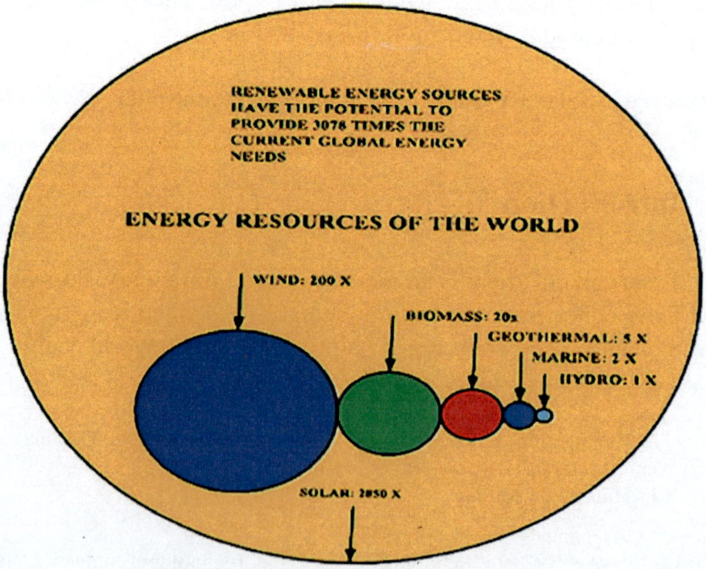

Fig. 1 Renewable energy resources of the world (Ellabban et al. [2])

2 Materials and Methods

An Arduino Atmega 2560 [7] is chosen to perform the functions as there are many pins on it and the experiment requires a lot of measuring and interfacing, it also has an ADC pin which is an analogue pin and can easily measure analogue data from the sensors required. A current sensor (ACS 712-20A) is used for this system. Its basic working principle is based on indirect sensing to determine the current passing through the wire. The voltage sensor is based on a simple voltage divider rule. The rating of the voltage sensor is any voltage below 25 V as the resistors used is to cater for this particular voltages. The Arduino Atmega 2560 is capable of changing is PWM frequencies to be up to 32 kHz easily with ease. This is another reason why the Arduino Atmega 2560 is chosen as the microcontroller used, so that the buck converter's MOSFET can be switched easily using the 5 V 32 kHz PWM pin. The complete system is shown in Fig. 2.

2.1 *Solar Tracking Design*

The solar 100 W PV panel was made to be imbalanced on the axle so that the auto gate motor can be the counter weight as shown in Fig. 3. The angle of the solar panel would be 45° to the horizontal axis to face either east or west with the 45° angle. The equipment used to move the structure of the solar tracker which is the Proton Wira power window motor which uses 12 V DC supply to control the turning of the solar panel based on the sensing of irradiance from the light dependent resistor circuit. The MD10C motor driver is used to control the motor. Pulse width modulation (PWM) is feed into the PWM pin of MD10C motor driver

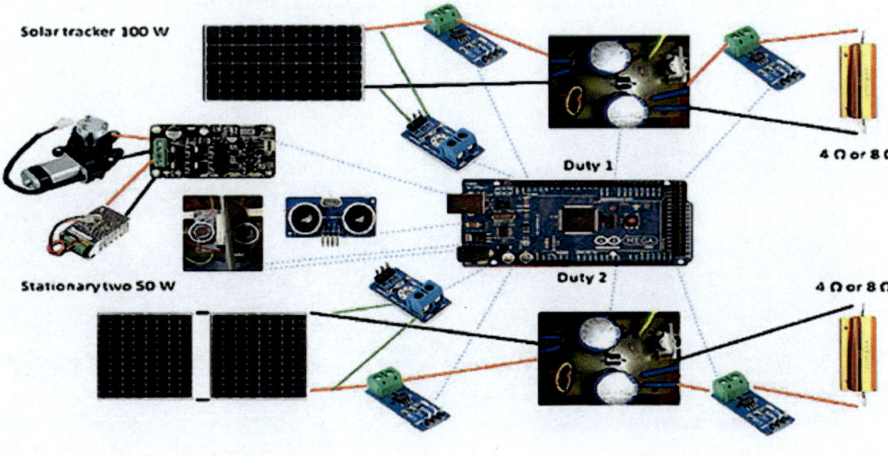

Fig. 2 Complete system of the diagram

to control the speed while DIR pin is used to control the direction and position of the solar panels. It is compared with a threshold and if it exceeds the threshold difference, it will move towards the direction of least resistance which means there is more irradiance.

2.2 Buck Converter Design

The buck converter [8] above works like a normal buck converter except the switching is controlled by the ground leg of the circuit whereby the output of the circuit is not grounded to the input of the circuit and is controlled by the IRF3205 MOSFET switch which is an n-type MOSFET which is required to switch the ground 'on' and 'off'. It is capable of turning 'off' and 'on' much higher currents than the gate pin of the MOSFET which is controlled by the Arduino Atmega 2560. The inductance value can be calculated as in Eq. (1):

$$L = \frac{D(Vs - Vo)R}{2f(Vo - Imin\ R)} \tag{1}$$

where D represents the duty cycle ratio of the PWM, Vs represents the voltage supply input, Vo represents the expected output voltage, R represents load resistance, f represents the switching frequency of the PWM and Imin represents the minimum current.

Based on the Eq. (1) above the assumptions are Vs = 19 V, Vo = 14 V, R = 8 or 4 Ω, f = 32 kHz, Minimum current = 0 A so that it will work in continuous mode. Even if minimum current increases, the inductance required will increase. Using the above assumptions the minimum inductance required is 44.64 μH. The inductance chosen is 100 μH that is capable of handling 5 A. The capacitance chosen at the input and output are the same which is 4700 μF. The input capacitor

Fig. 3 Solar panel placed in the tracker

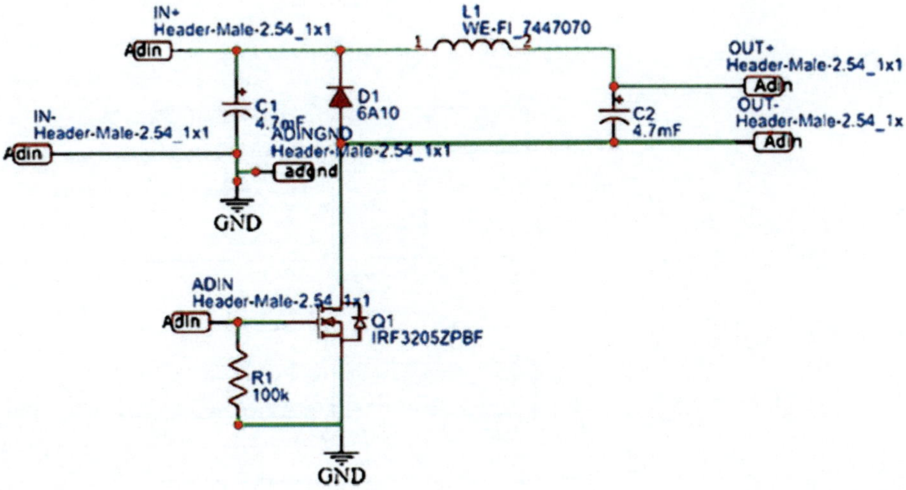

Fig. 4 Buck converter circuit design

is chosen to be so in order to reduce the amount of noise. It acts as a low pass filter such that high frequencies are reduced for a smoother waveform.

$$\text{Cout(min)} = \frac{\Delta \text{IL}}{8 \times \text{fs} \times \Delta \text{Vout}} \tag{2}$$

The input capacitor is chosen to be so in order Eq. (2) shows the rough value of output capacitance required. Whereby the estimated ripple current is determined by inductance shown in Eq. (3), switching frequency is 32 kHz and ripple output voltage is kept to a minimum so the value of output capacitance is increased. Thus, the value of the output capacitance is chosen to be 4700 µF. Equation (3) shows the rough value of inductor ripple current whereby Vinmax = 19 V, Vout = 14, fs = 32 kHz and L = 100 µH. This gives about 1.56(D) of ripple current depending on the duty cycle.

$$\Delta \text{IL} = \frac{(\text{Vinmax} - \text{Vout})\text{D}}{\text{fs} \times \text{L}} \tag{3}$$

2.3 *Maximum Power Point Tracking*

The Fig. 5 shows the flow chart of the MPPT being used. The flow chart is performed by the same Arduino Atmega that is being used to solar track the sun.

The perturb and observe method basically checks to its previous voltage and current(power) to determine whether duty cycle of the DC to DC converter has to change its operating voltage more to be more higher (shift up to right) or lesser duty

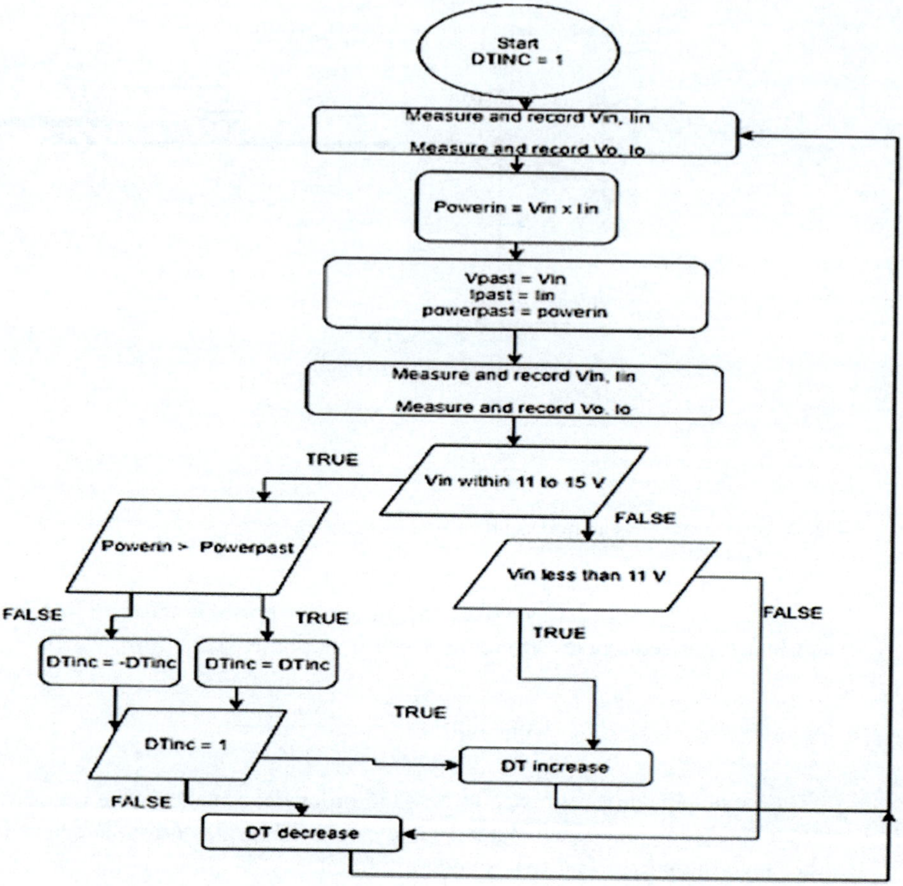

Fig. 5 Complete system of the diagram

cycle (shift down to left). It also checks the voltage output and makes sure that it is within the range of voltage it can be accepted. In this case between 10 and 15 V.

2.4 *R Characteristic Load*

The values of the resistors of 4 and 8 Ω has been chosen due to the solar panels R characteristics. This is defined by the maximum operating point of voltage and current of a specific solar panel. The R characteristic line is shown as Fig. 6. The experiment is repeated with the two different values of resistors to fit the two different R characteristics.

Using the equation to calculate the R characteristic, the R characteristic found for the 50 and 100 W panel are shown in Table 1.

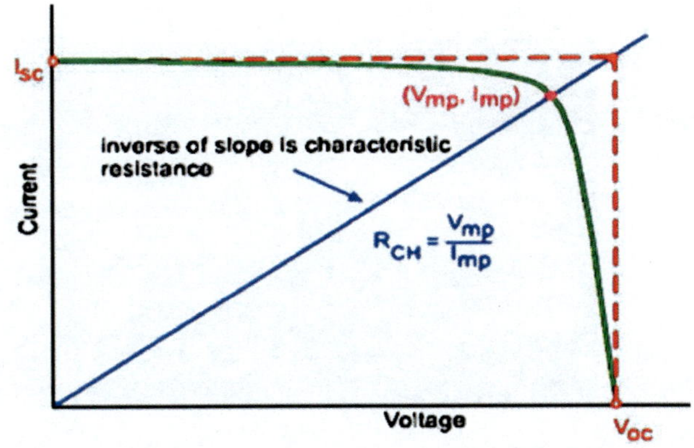

Fig. 6 R characteristic load

Table 1 R characteristic of 50 and 100 W

Solar PV (Watt)	Ohms
50	6.12
100	3.42

2.5 *Data Saving Method and Implementation*

All of the data that is being calculated and process will be recorded in Arduino's Serial port and will be kept to produce a graph for research. This is then saved into the computer used for the experiment by using an application call CooltermWin [9]. The application reads the Serial input from the Arduino microcontroller's port that is connected to the computer so that real time shown beside the data collected. The experiments of the stationary and solar tracking will be done together with the same value for the dummy load resistor which would be 4 Ω and 8 Ω respectively for two different experiments. The efficiency of the two systems will then be calculated and compared. The formula for efficiency will be determined by the power. Figure 7 shows the design to monitor a standard solar PV system in a residential household (pipe) for comparison with the single axis solar tracking system (wooden). The measured parameters that are most significant would be the output voltage and output current of the system. Since the irradiance is assumed to be the same since the experiments take place right beside each other, the temperature of the monocrystalline solar PV panels is assumed to be constant. The measurements that is to be recorded are the voltage and current of the solar panels before and after MPPT and with and without solar tracking.

The general program flow chart is shown below as Fig. 8. Noting that there is a 1 s delay reading of each set of input, output of solar tracker and stationary panels. This would allow rippling of the current and voltage to settle before the next reading, it is only an estimated safety measure so that the readings being read are correct.

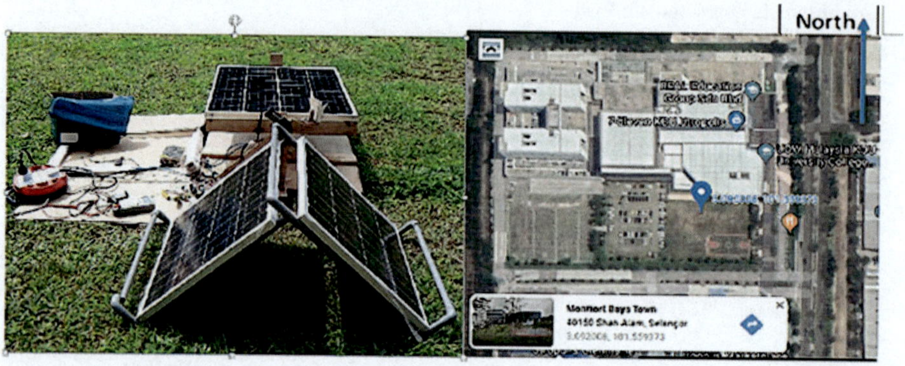

Fig. 7 Solar tracker and stationary panels facing the East-West direction on left the location of the experiment is at UOW KDU University College at the coordinates (3.092008, 101.559373)

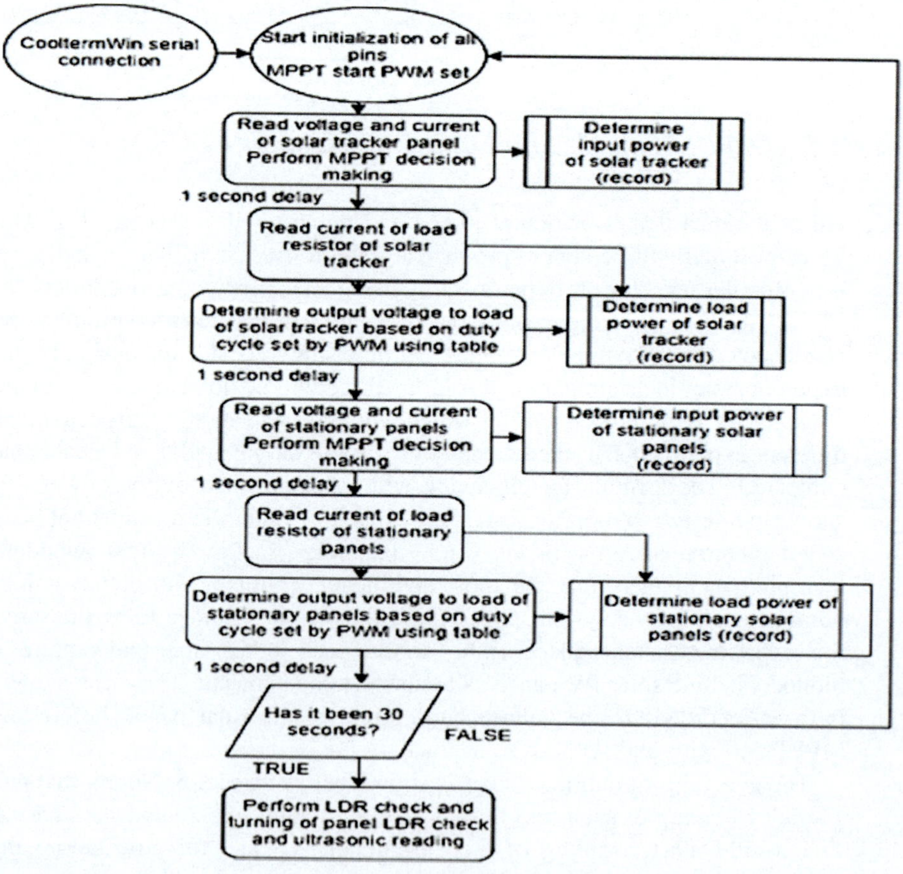

Fig. 8 General flow chart onto how both systems results and MPPT is determined

3 Results and Discussion

The overall design is made to monitor a standard solar PV system in a residential household (pipe) for comparison with the single axis solar tracking system (wooden). The measured parameters that are most significant would be the output voltage and output current of the system. Since the irradiance is assumed to be the same on each panel, since the experiments take place right beside each other, the temperature of the monocrystalline solar PV panels is assumed to be constant. The measurement that is to be recorded are the voltage and current of the solar panels before and after MPPT and with and without solar tracking. Irradiance represent one of the key factor that affect the Solar PV [3], which has been measured using solar meter (TES 1333R). As it has been shown in Fig. 9 the data was taken from 10 a.m. until 4 p.m. At 0 min irradiance was 1000 W/m^2 after 15 min it has drop to 75 W/m^2. Measurements of the power has been taken for the solar panels in the same time as shown in Figs. 10 and 11.

Time of experiment starts from 10:00 a.m. to 16:00 p.m. As shown in Fig. 10 star is solar tracker Ppanel. Circle is solar tracker Pload. Triangle is stationary panels in parallel Ppanel. Cross is stationary panels in parallel Pload.

It can be observed that the load power of the 100 W solar tracker is sometimes higher than the input power from the solar panel, this is due to tolerance of the current sensor, method of calculating output voltage and quick change in irradiance due to shading and dispersion. Since there is tolerance, the load power observed is very close to input power from the solar panel. It can also be seen that the two 50 W stationary solar trackers do not perform very efficiently after the buck converter when the load resistor used is 4 **Ω**. The actual R characteristic for a 50 W panel used in the experiment is about 6.12 **Ω**. Based on the results shown in Fig. 10 on the left, the power input from the solar tracker when compared to the stationary two 50 W panels is always higher. During noon, which is about 120 min in the graph shows that the power of the stationary panels is a little bit closer to the power of the solar tracker when compared to 10 a.m. or 0 min which is the very first few data's from the power input of the solar panels. It is observed from Fig. 12 on the left that the output power from the 100 W solar tracker is still higher than the two 50 W stationary panels output power.

As shown in Fig. 14 it can be clearly seen that although an 8 **Ω**, the load power of the 100 W solar tracker is still higher than the load power of the two 50 W stationary panels although the load resistor used is now closer to the R characteristics of the 50 W solar panel which is 6.12 **Ω** as shown in Fig. 13. It can be observed from Fig. 12 that is it harder to buck the voltage to be within voltage range set. Similarly, the system is also unable to keep the load voltage within the set range. The reason is also the same as there is an offset from the voltage sensor of −2 V which means that some of the higher voltages is within voltage range and the lower one is not within the range. Another cause would be the size of the output capacitor, which is 4700 μF, there is charging and discharging time of the capacitor

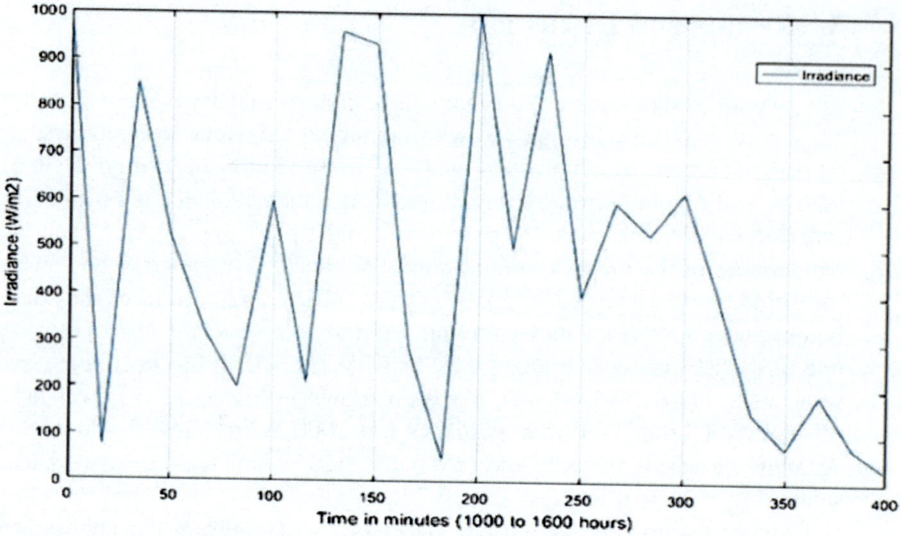

Fig. 9 Irradiance data

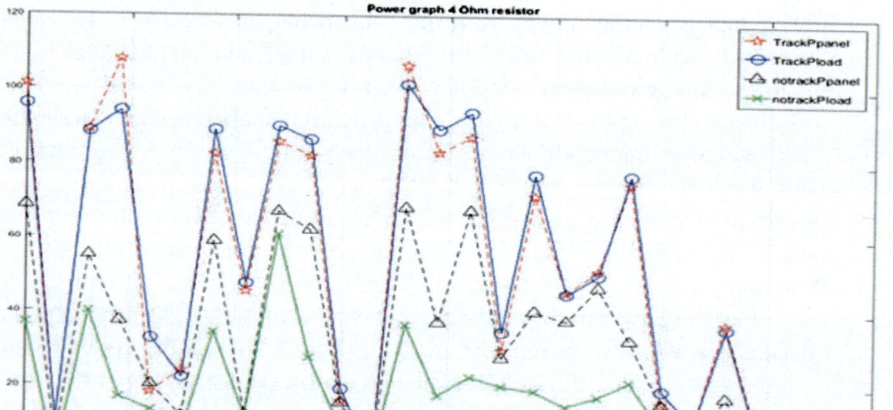

Fig. 10 Comparison of the output power of PV cell and load output power, of 100 W solar tracker and two 50 W stationary panels connected in parallel, using 4 ohms

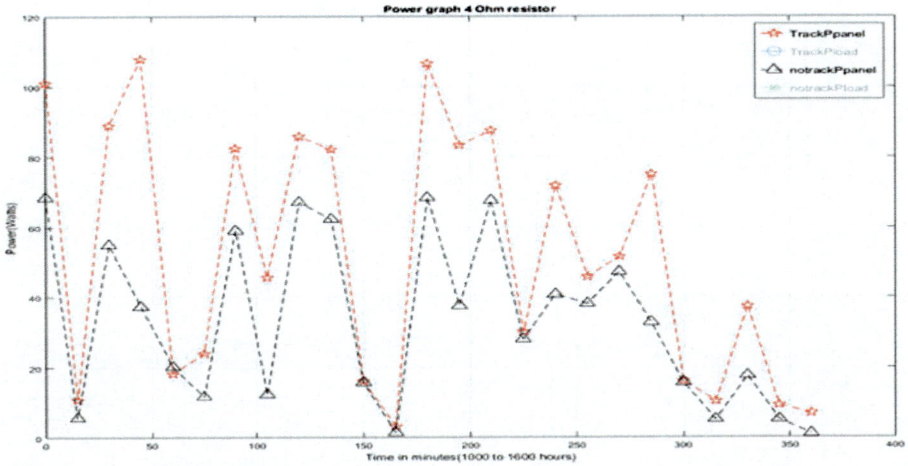

Fig. 11 Comparison of the output power with and without tracker, of 100 W solar tracker and two 50 W stationary panels connected in parallel, using 4 ohms

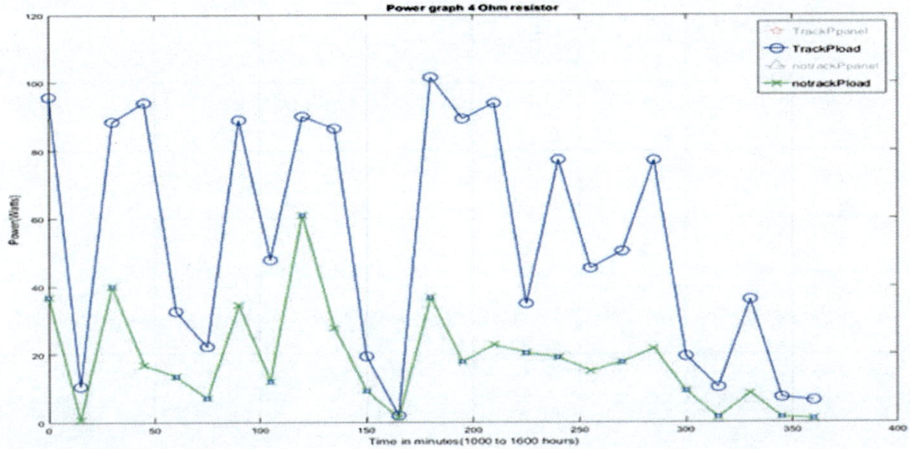

Fig. 12 Comparison of the load output power with and without tracker, of 100 W solar tracker and two 50 W stationary panels connected in parallel, using 4 ohms

which may affect the voltage data being collected whereby it takes longer time to change voltage with the buck converter. Since the output load resistance is very small, the discharge time is also reduced therefore not affecting the result very much. Another reason why the load voltage can stay within the voltage range set is because of the climate in Malaysia whereby it is the monsoon season during November.

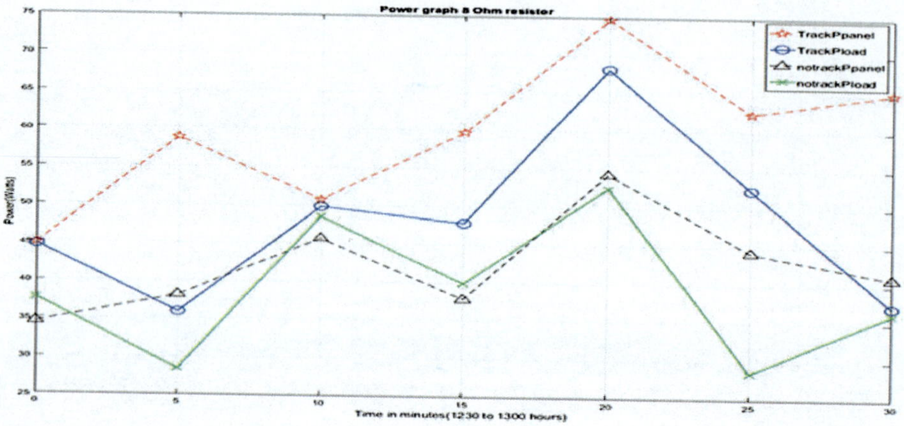

Fig. 13 Comparison of the output power of PV cell and load output power, of 100 W solar tracker and two 50 W stationary panels connected in parallel, using 8 ohms

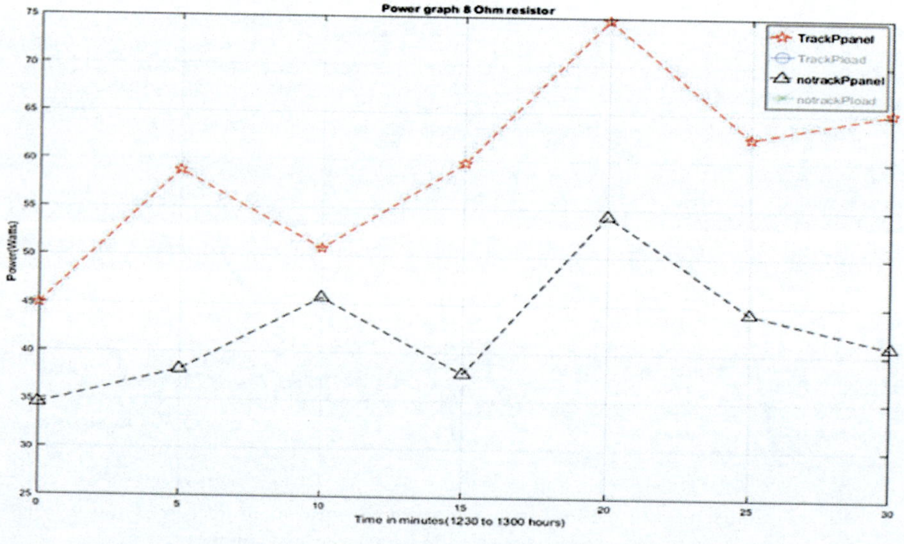

Fig. 14 Comparison of the output power with and without tracker, of 100 W solar tracker and two 50 W stationary panels connected in parallel, using 8 ohms

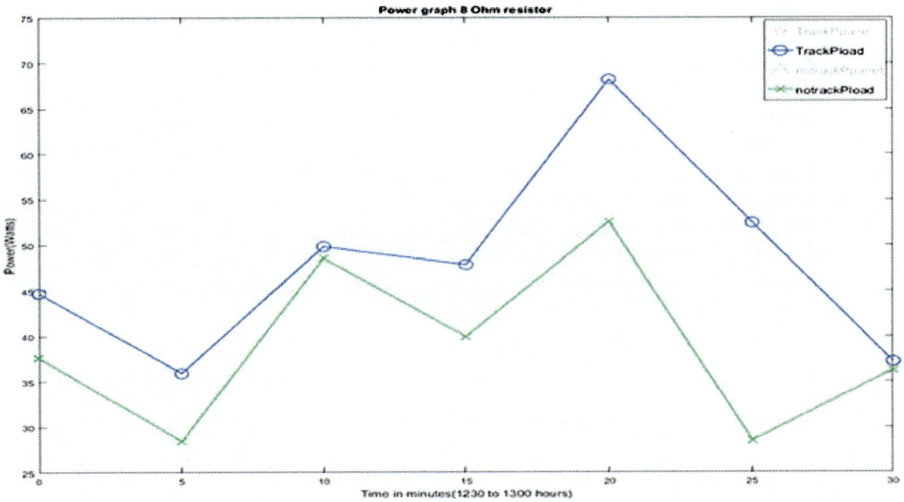

Fig. 15 Comparison of the load output power with and without tracker, of 100 W solar tracker and two 50 W stationary panels connected in parallel, using 8 ohms

4 Conclusions

The experiment was carried out at the same location and same time using two different load resistors and the results show that solar tracking has better performance than no solar tracking even with different load resistors for the 50 W solar panel and 100 W solar panel. The results show that the solar tracking system with MPPT can further optimize the output power to the load by further taking into consideration the R characteristic of a solar panel. By using energy efficient components such as the worm motor that doesn't use any power to keep the solar panel in place and further improvements by using gearing, the systems efficiency is further increased.

References

1. Doman L (2017) EIA projects 28% increase in world energy use by 2040, 14 Sept 2017. Available https://www.eia.gov/todayinenergy/detail.php?id=32912
2. Ellabban O, Abu-Rub H, Blaabjerg F (2014) Renewable energy resources, current status, future prospects and their enabling technology. In: Renewable and sustainable energy reviews, vol 39, Nov 2014, pp 748–764
3. MohamedZain AO, Chua HS, Yap KM, Al-Talib AAM (2020) Outdoor experimental comparison between PV cell and combined system PV-TEG. Sci Int (Lahore) 32(3):279–283
4. Mohamedzain OA, Govinda S, Chua HS, Al-Talib AAM (2020) Experimental investigations of the effect of temperature on power in a combined photovoltaic cell and thermo-electric. Lecture notes electrical engineering, vol 619, Nov, pp 199–210

5. Albert FYC, Mason CHS, Kiing CKJ, Ee KS, Chan KW (2014) Remotely operated solar-powered mobile metal detector robot. Procedia Comput Sci 42:232–239
6. Kotrov OI, Smulsky JJ (2014) New computing algorithm of the earth's insolation. Appl Phys Res 6(4):56–82
7. Gusa RF, Sunanda W, Dinata I, Handayani TP (2018) Monitoring system for solar panel using smartphone based on microcontroller. In: Proceedings—2018 2nd international conference on green energy and applications ICGEA 2018, pp 79–82
8. Baharudin N, Mansur T, Nizar TM, Hamid F, Ali R, Irwanto M (2018) Performance analysis of DC–DC buck converter for renewable energy application. J Phys: Conf Ser 1019:012020
9. Mason CHS, William Y, Albert FYC (2014) Design and implementation of sub-GHz wireless light switch with integrated Wi-Fi. In: 2014 IEEE international conference on communication networks and satellite, pp 34–38. IEEE

Dielectric Properties Measurement of Agarwood Using Vector Network Analyzer for Frequency 200 MHz Until 1 GHz

Nurfarahin Ishak, Chua King Lee, and Siti Zarina Mohd Muji

Abstract Non-destructive technique has potential to be used in agricultural for example to inspect the formation of agarwood. However, to build a system using the non-destructive technique need to know the dielectric properties itself. The aim of this paper is to measure the dielectric properties of agarwood by using Vector Network Analyzer (VNA). Measurement was conducted at frequency 200 MHz until 1 GHz with open-ended coaxial probe method. Both relative permittivity and conductivity were affected by frequency, moisture, and temperature. The most significant influence of dielectric properties was observed between these ranges of frequencies. The relative permittivity decreases with increasing the frequency and conductivity showed maximum and minimum values under various conditions.

Keywords VNA · Relative permittivity · Conductivity

1 Introduction

Aquilaria Malaccensis also called (agarwood) is a species of plant that can be found in Malaysia. Agarwood is commonly known as Gaharu which produced fragrant resinous wood. The formation of resin occurs as a response to fungus infection in the stem of the trees. Generally, the resin raises the mass and density of the infected wood and shifts the colour of the bark to black. [1]. Agarwood is high valued wood for extensive uses such as perfumes, incense, and medicine. China is famous produced agarwood products such as carminative, anti-emetic effects and sedative

N. Ishak · C. K. Lee (✉) · S. Z. M. Muji
Faculty of Electrical and Electronic Engineering, Universiti Tun Hussein Onn Malaysia, Parit Raja, Malaysia
e-mail: chua@uthm.edu.my

N. Ishak
e-mail: farah.ishak25@gmail.com

S. Z. M. Muji
e-mail: szarina@uthm.edu.my

© The Author(s), under exclusive license to Springer Nature Singapore Pte Ltd. 2021

Z. Zakaria and S. S. Emamian, *Advances in Electrical and Electronic Engineering and Computer Science*, Lecture Notes in Electrical Engineering 741,
https://doi.org/10.1007/978-981-33-6490-5_4

medication [1]. The natural forming of high-quality agarwood takes several years or might be decades until the tree is affected by any extrinsic causes, such as lightning strikes, animal weeding, insect attacks or microbial invasions [2]. Due to the economic value, the agarwood has been exposed to the exploitation by the agarwood hunter. The agarwood farmer uses the inoculants which act as a booster and inject into the tree to expedite the agarwood formation. Compare to the traditional wounding methodologies, the chemical inducers seem to be able to rapid the agarwood formation. The inoculants are containing active compounds made from the microorganism or chemical concoctions.

Regarding understanding the electrical properties of wood, instrumental and industrial use, wood's dielectric properties are beneficial. Several researchers have recently been investigating the use high-frequency and microwave approaches to increase reliability of energy and end-product quality. The interaction between the alternating electromagnetic field and wood allows different properties of this substance to be elucidated. There are two components in the electromagnetic field: electrical and electromagnetic fields. Such different compounds' effect on wood. The magnetic field effect on wood is insignificant, however the impact of the electrical field in wood may occur since the interference with the magnetic field findings in the development of electrical currents in the material [3].

For this experiment, coaxial probe open-ended was chosen because do not need sample machining, and quick preparation of the sample. After measurement, a wide number of samples of dielectric properties may be calculated regularly over a limited period of time. The measure can then be carried out in a temperature-controlled environment.

2 Definition and Principles

The measurement of dielectric properties includes of the materials' relative permittivity (εr), relative permeability (μr) measurement and conductivity (σ). The permittivity properties consist of a real section and an imaginary section. Permittivity, also called as dielectric constant, is the calculation of the sum of energy from the exterior electric energy contained in the material. For lossless materials, the imaginary element is zero which is often regarded as the loss factor. It is a representation of the level of energy degradation of the substance owing to an external electrical field. The term tan δ is named loss tangent and reflects the connection between the hypothetical component and the real element of the permittivity. Terms such as tangent loss, dissipation factor, or loss factor also apply to the loss tangent [3]. The permeability is the measure of the resistance of a material against the formation of a magnetic field. The conductivity is the measure of the ease at which an electric charge or heat can pass through a material.

The dielectric parameters of wood can be influenced by several factors. Several studies have been published on the relationships between dielectric wood parameters and relevant material properties such as porosity rate, pore scale, pore allocation, moisture levels and density [4]. It is also correlated with the microstructure of wood, the interaction mechanisms between the wood and water, and the behavior of wood under a high-frequency electromagnetic field [5]. It has been observed that the dielectric properties can be affected by the frequency [4–6]. With increasing frequency and loss coefficient increases, relative permittivity decreases with increasing frequency for all species of wood and fiber orientation [4]. As dielectric properties studies became more complex, many factors such as temperature, structural direction, and density are considered with the findings showed the other factors do had something significant effect wood's dielectric behavior [7].

3 Experiment Setups

3.1 Test Sample Preparation

A cylindrical of agarwood samples were chosen for experiment. Then, it cut into three samples with the length 15 and 59 cm in diameter. Each of the samples has different size agarwood formation inside because it is formed scattered. The measurement reading is taking until nine (9) points in horizontal condition for precise and accuracy. The points were taken surrounding the samples. However, the structure of the test samples is not flat and it was an obstacle to get an accurate reading because the air gap has occurred and only reflection measurement is available.

There are several steps has been taken to prevent reading error and overcome the solution for air gap:

- Take the air reading before starting the experiment. Dielectric constant value at room temperature (25 °C, or 77 °F) is 1.00059. This value is compared to the reading value and make sure the reading place have no air gap.
- Take a few times reading at one point.
- Find the best point which has a flat area to place a probe.

The reaction of relative permittivity and conductivity toward frequency, temperature and moisture content are recorded.

3.2 Apparatus Setups

Calibrations are necessary before experimenting. The purpose of the calibration operation is to establish a relationship between the measured complex reflection coefficient and the predicted one. This method enables all post-calibration measurement information to be fixed. If carried out properly, a good calibration procedure results in reliable readings. Figure 1 below shows the experiment setup equipments.

Fig. 1 The experiment setup

The copper is well known as good conductivity. During calibration and the conductivity of copper used in the validation step equally—indeed, the same types of reference can be used either for calibration or validation as in Fig. 2.

The temperature of the load fluid must be preserved and controlled during the calibration phase, as the dielectric parameters are temperature-dependent. Liquid permittivity varies by up to 2.2% per degree Celsius. Measurement of deionized water or any normal liquid as a load of calibration may be conducted at room temperature or any set temperature. If the temperature of the liquid differs from the temperature of the probe, it is suggested to wait for the temperature to stabilize before continuing with the measurement as in Fig. 3.

3.3 Testing Open-Ended Coaxial Probe

The coaxial probe technique is ideal for liquid and semi-solid (powder) materials. The method is simple, reliable, non-destructive with one measurement. In this mode, the sample is pushed contrary a specimen or dissolved in the solvent and the coefficient of reflection is measured and used to determine dielectric properties. Dielectric constant, loss tangent and conductivity data were taken by Rohde & Schwarz Vector Network Analyzer (VNA). The frequency range is used from 200 MHz to 1 GHz. The system composed of an open-ended coaxial probe DAK 3.5 with 3.5 mm diameter and fully computer interfaced as in Fig. 4. The data from open and short response of coefficient was measured by the VNA system.

Next, the probe device is calibrated in such a way that the reflection coefficient measurement is apply to the aperture plane of the probe. This can be achieved by using reference liquids at the open end of the sample for precise calibration. The specified fluid is used as a comparison norm and must be a fluid with proven dielectric properties as in Fig. 3.

Fig. 2 Calibration with copper plate

Fig. 3 Calibration with water

While testing the specimens the coaxial probe must have a strong contact with the specimens. Any air gap between specimen and probe would affect the readings [8]. As microwaves are applied to a substance, some of the energy is being obscured, some of it transferred via the surface and some of the part is consumed by the latter level. The proportions of energy in these three classes were described as dielectric

Fig. 4 DAK 3.5 coaxial probe

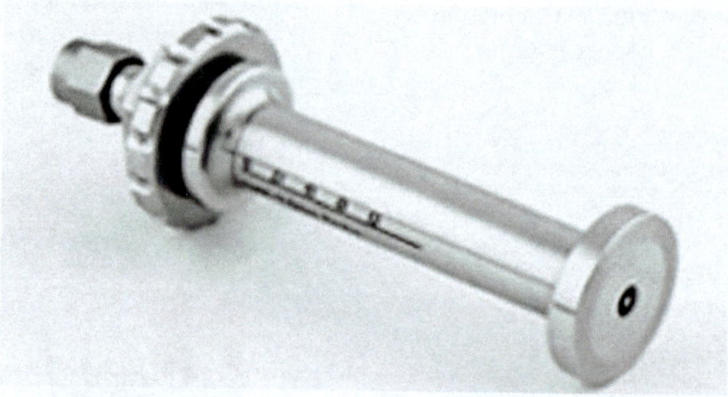

properties [9]. The basic electrical properties from which the encounters take place are represented is the relative permittivity of the material ε* [9, 10]. The mathematical expressed as:

$$\varepsilon^* = \varepsilon' - j\varepsilon'' \tag{1}$$

where

ε′ = dielectric constant, and
ε″ = dielectric loss factor.

The dielectric loss factor is a function of conductivity and frequency:

$$\varepsilon'' = \sigma/2\Pi f \tag{2}$$

$$\varepsilon = \varepsilon' - j\sigma/\omega \tag{3}$$

where

$\omega = 2\prod f$
σ = conductivity.

4 Dependence of Dielectric Properties

4.1 Frequency Dependence

Polarization as a consequence of the alignment with the applied electrical molecule field with persistent dipole moments is a significant factor responding to the reliance on frequency of certain dielectric properties of materials [7, 11]. Different kinds of the dipole would have different times of relaxation. Therefore, as the frequency rises, the permittivity declines in phases and the dielectric loss peaks.

Dipolar water molecules are lined up in the field at low frequencies and the relative permittivity received approximately 80. The permittivity decreases in a phase at higher frequencies, and the power factor has a limit [12]. Previously reported was the effect of the measurement conditions such as temperature, frequency, and orientation of the electrical field concerning the wood structure [4].

4.2 Temperature Dependence

Relative permittivity can increase with rising temperatures in a dispersion area, while the loss factor can change depending on whether the frequency of operation is different from frequency of relaxation [11].

4.3 Density and Moisture Dependence

In the surrounding water, the undefinable regions are more permeable as the amount of moisture rises. It is reasonable to presume that ions will move through the amorphous regions under the influence of the external electrical field and concentrate at the edges of the crystals until the inner field of separated ion charges is equivalent to the applied field, or until the external field is inverted. [7].

5 Result and Discussion

This experiment was tested at the three *Aquilaria* samples with agarwood formation. Each sample was labeled as A, B and C as in Fig. 5. Nine points readings were taken for each sample. The data has precise by using the standard deviation and mean calculation. From the statistic graph value shows the patterns movement in increment of frequencies. The size and location of agarwood inside the samples are different. This is because of the formation of agarwood are scattered. Therefore, size agarwood formation is influenced by the reading of the dielectric. The different size formation of agarwood inside the sample giving a different measurement reading. Sample A have a least formation gave high reading while sample C have large formation gave low reading of permittivity.

The relative permittivity indicates the molecule's tendency to polarize within the electrical field and the loss coefficient indicates the material ability to convert electromagnetic energy to heat at a specific frequency and temperature. At higher frequencies, relative permittivity is found to decrease further effects are attributed to the involvement of electrical polarization. Many additions are negligible and even decline as polarization will not arise instantaneously owing to the strength of the electrical field [12]. All forms of polarization apply at low frequencies. As

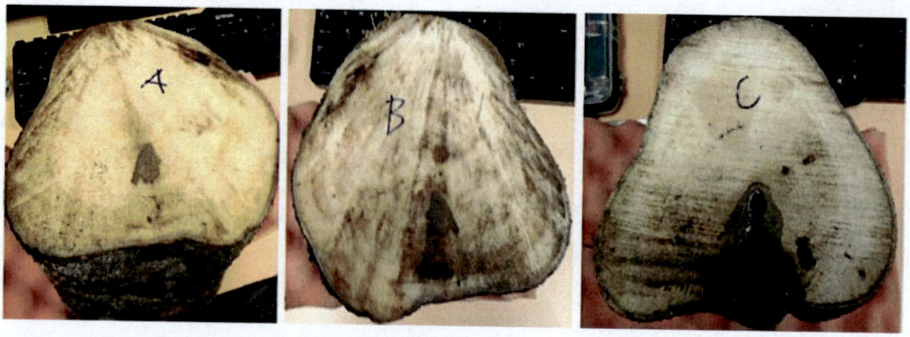

Fig. 5 The fresh *Aquilaria* sample with agarwood

frequency increases, those with long time of relaxation times cease to react, and thus the relative permittivity decreased [5, 6].

The relative permittivity and conductivity of Aquilaria samples have been measured by using VNA with a frequency range from 200 MHz to 1 GHz. High values of the relative permittivity (ε') and conductivity (σ) found due to moisture content. Observation was found that temperature and frequency indicating the conduction mechanism, where the thermal conductivity of wood comparatively poor owing to the porosity of timber as wood is a very complex material with low permeability and thermal conductivity. Thermal conductivity increases as the wood density increases. The higher the amount of moisture was, the greater the frequency impact on the factor of loss.

The graph in Figs. 6 and 7 below shows the linear regression of relative permittivity and conductivity for samples A, B, and C. These graphs are the correlation between average relative permittivity and conductivity versus frequency. The previous researcher showed the example approach to determine their result whether to be reliable by the interpreting by correlation coefficient as in Table 1 [13]. Linear regression (R) is a common technique for statistical data analysis. It is used to determine the extent to which a consistent relationship exists between a dependent variable and one or more independent variables. By referring to the Table 1 correlation coefficient shows that the value regression of relative permittivity in the graph is moderate correlation where sample A, R2 = 0.5704 while sampling B, R2 = 0.5474 and sample C, R2 = 0.1939. The R2 value for sample C is low because of the sample condition where has a hole inside and the agarwood formation is not solid. The accuracy reading distracted by the air gap. Then, for the regression of conductivity result shows that near to strong correlation. For sample A, R2 = 0.9019, sample B, R2 = 0.5031 and sample C, R2 = 0.7981. This correlation results, depending on the initial results of the measurement. As in the discussion, many factors are influencing the readings such as moisture, temperature, and frequency. However, to achieve a strong correlation need to minimize the factors that interfere during the measurement reading.

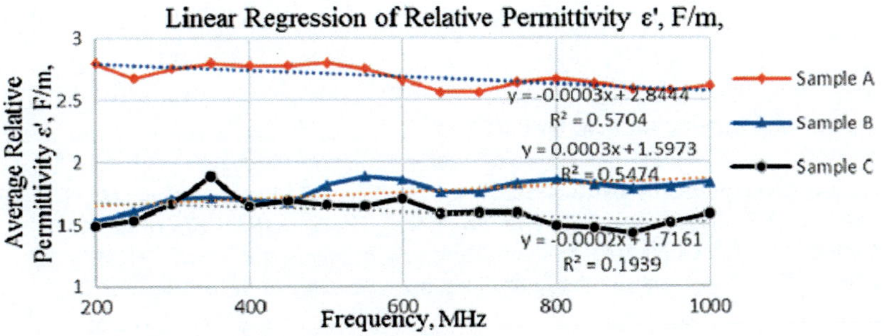

Fig. 6 The linear regression of relative permittivity

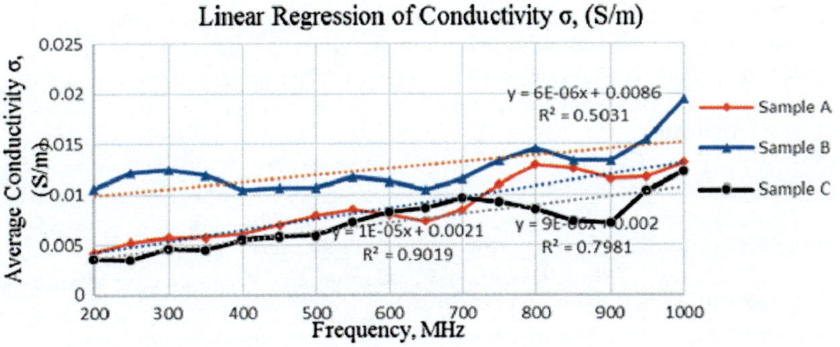

Fig. 7 The linear regression of conductivity

Table 1 Correlation coefficient interpreting table [13]

Correlation coefficient	Interpretation
0.00–0.001	Negligible correlation
0.10–0.39	Weak correlation
0.40–0.69	Moderate correlation
0.70–0.89	Strong correlation
0.90–1.00	Very strong correlation

6 Conclusion

The measurement dielectric properties of agarwood with used the open-ended coaxial probe method shows that the relative permittivity data slightly decreasing while the frequency is increased opposite to the conductivity, the measurements data is increasing while the frequency is increased. The measurement data are massively inclined by frequency, temperature, and amount of moisture. However, fatal error such as air gap during measurement processed have also given effect on dielectric reading.

The relative permittivity measurement at frequency 650 MHz until 700 MHz for all samples shows the flatted readings because of relaxation times cease to react. The dielectric properties data of agarwood provided would fulfill all of the needs for potential high-frequency research.

Acknowledgements This research was supported by Fundamental Research GRANT Scheme (FRGS) research Vote K087 (Ref: FRGS/1/2018/TK04/UTHM/02/28). Thank you Ministry of Education (MOE) and Research Management Centre (RMC) of Universiti Tun Hussein Onn Malaysia for the financial support to achieve the research goal.

References

1. Chen X et al (2018) Trunk surface agarwood-inducing technique with *Rigidoporus vinctus*: an efficient novel method for agarwood production. PLoS ONE 13(6):1–13
2. Chen X et al (2017) Agarwood formation induced by fermentation liquid of *Lasiodiplodia theobromae*, the dominating fungus in wounded wood of *Aquilaria sinensis*. Curr Microbiol 74(4):460–468
3. Torgovnikov GI (1993) Dielectric properties of wood and wood-based materials. Springer-Verlag
4. Pento K, Łuczycka D, Wysoczanski T (2017) Dielectric properties of selected wood species in Poland. Wood Res 62(5):727–736
5. He X et al (2017) Study on dielectric properties of poplar wood over an ultra-wide frequency range. BioResources 12(3):5984–5995
6. Varada Rajulu KC, Mohanty BN (2016) Dielectric and conductivity properties of some wood composites. Int J Eng Technol 8:51–60
7. James WL (1975) Dielectric properties of wood and hardboard variation with temperature, frequency, moisture content, and grain orientation
8. Riminesi C, Olmi R, Bini M, Ignesti A (2000) Dielectric of properties of wood from 2 to 3 GHz. J Microw Power Electromagn Energy 35(3)
9. Venkatesh MS, Raghavan GS (2005) An overview of dielectric properties measuring techniques. Can Biosyst Eng 47
10. Norimoto M (1976) Dielectric properties of wood
11. Nelson SO, Kraszewski AW (1990) Dielectric properties of materials and measurement techniques. Dry Technol 8(5):1123–1142
12. Morton WE, Hearle JWS (2008) Dielectric properties. In: Physical properties of textile fibres, pp 625–642
13. Schober P, Schwarte LA (2018) Correlation coefficients: appropriate use and interpretation. Anesth Analg 126(5):1763–1768

Energy Consumption and Cost Estimation Considering Renewable Energy Sources in Integrated Next-Generation Wireless Access Network

Check for updates

Arnidza Ramli, Abubakar Abdulkadir, Nadiatulhuda Zulkifli, and Adam Wong Yoon Khang

Abstract Energy consumption in telecommunication industries is a major challenge for the network providers where most of the power consumed equipments are from the wireless access network and the optical network transmission. The non-renewable energy are the main sources of electricity such as diesel, burning fossil fuel and only a small percentage of energy comes from clean renewable energy source such as solar, wind, biomass and tidal power. Therefore, this research work is aimed to estimate the precise power consumption of an integrated optical-wireless access networks by characterizing the most power demanding equipment and providing an alternative source of a clean renewable energy to power them. The main advantage of this technique is the simplicity in which the power consumption can simply be modelled by using Matlab and to mitigates the GHG emissions of the system by hybridizing renewable energy which are solar and wind into existing energy sources and simulated using Homer Pro. The obtained results were based on Skudai-Malaysia location where 77% reduction of carbon emissions through renewable integration and 56% operational cost were achieved.

Keywords Next-Generation-Passive optical network · Operational cost · Power consumption · Renewable energy · Wireless access network

---

A. Ramli (✉) · A. Abdulkadir · N. Zulkifli
Lightwave Communication Research Group, Faculty of Engineering, Universiti Teknologi Malaysia UTM, 81310 Johor Bahru, Malaysia
e-mail: arnidza@fke.utm.my

A. W. Y. Khang
Broadband & Networking Research Group (BBNET), Center for Telecommunication Research and Innovation, Fakulti Teknologi Kejuruteraan Elektrik Dan Elektronik (FTKEE), UTEM, Hang Tuah Jaya, Melaka, Malaysia

© The Author(s), under exclusive license to Springer Nature Singapore Pte Ltd. 2021

Z. Zakaria and S. S. Emamian, *Advances in Electrical and Electronic Engineering and Computer Science*, Lecture Notes in Electrical Engineering 741,
https://doi.org/10.1007/978-981-33-6490-5_5

1 Introduction

The power consumption in mobile communication industries are the major concern. The integration of next generation-passive optical network (NG-PON) and Wireless Access Network (WAN) provides high speed broadband network, but energy consumption increases due to network overloads by additional number of computers connected to the base stations (eNBs) network [1]. The integrated NG-PON-WAN systems is powered by diesel generators which produces green-house gases (GHG) emission (environmental impact) [2]. The diesel generator releases polluted air in forms of Carbon dioxide, carbon monoxide, ammonia, nitrous oxides, sulfur dioxide, chlorofluorocarbon and methane are the major contributors of global warming causes genetic change, depletion of ozone layer, cancer and acid rain [3]. Therefore, this paper proposed a method to estimate the energy consumption as well as cost by integrating renewable energy source to power all the major hungry components for the entire architecture.

2 Related Works

Almost a decade, significant achievement has been made by hybridizing renewable and conventional energy sources into a smart grid to succeed in supplying electricity to the base station (BS) sites. In [4], the authors presented an analytical model and evaluated the power consumption based on the integration of PON-Wireless networks. However, the valuation of energy consumed by the components for such networks have been realized significantly. The analytical model used for measuring the energy consumption of the network elements was derived from the datasheets reported in [1, 4]. Therefore, the overall power consumption model for the integrated networks was expressed as:

$$P_{IOB} = \sum_{i=1}^{M} P_i^{OLT} \times \sum_{i=1}^{N} \left(P_i^{ONU} + P_i^{FBS}\right) \tag{1}$$

where M and N are the number of an optical line terminal, OLT and the power integrated optical-base station, P_{IOB}, the P_i^{OLT}, P_i^{ONU}, and P_i^{FBS} are the power consumed by the OLT, optical network unit (ONU), and Femto-Base Stations (FBS) respectively. The individual power consumption models for different network components was reported in [4]. The power consumption mostly depends on the traffic load and distance (or site factors). However, the power consumption of the end-user equipment's was not accounted such as laptops and smartphones.

Authors in [5], proposed the feasibility study and simulation of an integrated Solar Photovoltaic with battery storage to power some selected mobile BSs within the Soshanguve in Pretoria, South Africa. The proposed study was aimed at the possibility of implementing a hybrid power system (HPS) for cellular BSs and analyzes the economic and environmental impact of the study against the diesel generator set as the main source of electricity to BSs. The total selected BSs power was modelled [5] as:

$$P = \sum_{n=1}^{N} P_n \cdot \frac{T_n}{24\,\mathrm{h}} \quad (2)$$

where P represents the total DC load power, P_n and $\mathrm{T_n}$ are the components power consumption is the run time (n = 1, 2, 3, …). However, the proposed technique used to consider the solar PVs architecture for a period of 10 years only while the result obtained was able to achieve reduction in GHG emission and operational cost as compared to the diesel genset, but still for 10 years project, the total net present cost (NPC) was at $255,812.00.

3 Methodology

3.1 *Optical-Wireless Access Network Architecture*

Figure 1 shows the integrated NG-PON-WAN architecture considered in this paper. The architecture can be divided into four parts, three of which are considered as power hungry components and the units are, the OLT, ONU, eNBs and another one part passive splitter node. The maximum fiber distance between OLT and ONU are typically 40 km, but is capable to reach up to 60 km, depends on the bandwidth capacity and service requirements. The ONU is connected to OLT through the passive splitter with splitting ratio up to 1:256 considered in this work. The eNBs are located at customer premises equipment (CPE) and they are long term evolution (LTE) base stations connected to ONUs that offers internet access to the mobile users.

3.2 *Network Power Consumption Model*

The power consumption (PC) model of an integrated NG-PON-WAN architecture shown in Fig. 1 can be expressed as [6]:

$$P_{overall} = \frac{3}{2}(P_{SN} + P_{EN}) + P_{CPE} \quad (3)$$

where $P_{overall}$, P_{SN}, P_{EN} and P_{CPE} are the overall PC of an integrated NGPON-WAN, the power consumed at the splitting ratio SN (typically is zero),

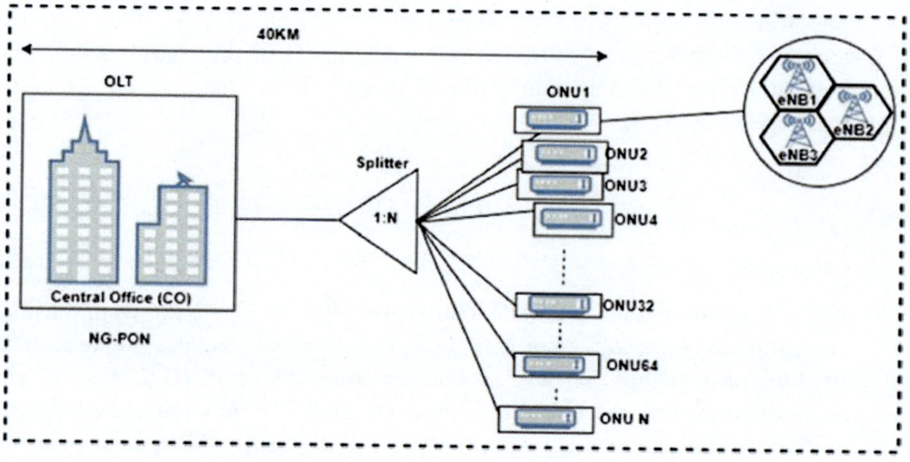

Fig. 1 NG-PON-WAN architecture

because it's a passive component, the power at edge node EN and plus the power consumed at the CPE node. The constant fraction number, $\frac{3}{2}$ is a multiplier power factor usage for the effective cooling and the uninterrupted power system (UPS), located at the network site.

The NG-PON described as an OLT equipment located in central office is referring to as the PC at the P_{EN} and can be modelled as [6]:

$$P_{EN} = P_{ONU} = P_{idle} + \left(\frac{P_{idle} + P_{\max}}{C_{\max}}\right) \times C \quad (4)$$

where P_{ONU} is the PC at ONU, $P_{idle}, P_{\max}$, are the energy consumed at maximum peak load and at no-load measured in watts (W), while $C_{\max}$ and C are the maximum traffic load and the data rate, all are measured in bit per second (bps). The parameters for both the OLT and the ONU equipments used in this work was based on "Cisco ME4600 series" and the simulation parameters selected from its datasheet [7, 8].

The International Telecommunication Union-Telecommunication (ITU-T G.989.1) recommendations of the NG-PON provided the symmetric peak data-rate, C for downlink and uplink as 10 Gbps/10 Gbps. Therefore, both the OLT and ONU equipments (Cisco ME4600 series) satisfied the requirements of the maximum load traffic at 5 Gbps and the data rate, C at 3 Gbps was selected because only remote or less populated are considered in this work. The $P_{\max}$ of both the OLT and the ONU (ME4624-ONT-RGW) are <545 W and 15 W accumulated energy consumption reported in datasheet [7, 8]. In an idle situation, the power consumed at less traffic (no-load) is 90% of the $P_{\max}$ as reported in [7]. The PC of an OLT and ONU are summarized on Table 1.

Table 1 Power consumption parameters [7, 8]

	P_{EN} $(ME46XX{-}OLT)$	P_{ONU} $(ME4624{-}ONT)$
$P_{\max}$	540 W	15 W
P_{idle}	486 W	13.5 W
$C_{\max}$	5 Gbps	5 Gbps
C	3 Gbps	3 Gbps

The LTE-eNB described as the major power-hungry equipment that contains the power amplifier (PA) known as radio-remote head (RRH), radio frequency (RF) unit, the baseband (BB) unit, the air-conditions (ACs) used for cooling system, security lighting (SL). The above mentioned component's PC was calculated and reported [9, 10]. The LTE-eNB power model can be expressed as [11]:

$$P_{eNB} = \frac{P_{PA} + P_{BB} + P_{RF}}{((\sigma_{cool}) \times (\sigma_{ms}) \times (\sigma_{dc}))} \times N_{sector} \tag{5}$$

The LTE-eNB PC of the following components; the RRH, the integrated RF and BB devices considered in this work are summarized in Table 2:

The P_{eNB} refer to PC of an LTE-eNBs, P_{PA}, P_{BB}, P_{RF} are the PC for power amplifier, baseband and radio frequency units measured in watts. Other symbols of the denominator in Eq. (5) are losses due the cooling system (σ_{cool}), the main-supply (σ_{ms}) and dc-to-dc (σ_{dc}) conversion losses [11]. The N_{sector} denotes to the number of base station sectors used as depicted on Fig. 1. The PC at the CPE node refers to as the summation of single ONU and the eNBs. Therefore, mathematically, P_{CPE} expressed as [4]:

$$P_{CPE} = P_{eNB} + P_{ONU} \tag{6}$$

The entire energy consumed by an P_{CPE} was calculated as 6589 W, that is including the two pieces of 2-horse power AC's each per sector used for network equipment cooling, two security light of 200 W each used for exposing a trespasser within the base station.

Table 2 LTE-eNB PC parameters [11]

Components	Values
P_{PA}	185 W
$P_{RF} + P_{BB}$	55 W
σ_{cool}	9%
σ_{ms}	7%
σ_{dc}	6%
N_{sector}	3

Table 3 Required bandwidth for FTTH services

Applications/Services	Required Bandwidth (Mbps)
Additional TV channels	16
Internet downloads	10
High-definition TV	8
Standard definition television	3
Music on demand	2
Multimedia	2
Online gaming	1
Voice over IP (VoIP)	1
Video conferencing	1
Total	44

3.3 *FTTH Scenario*

Based on the recommendations from the ITU-T G.989.1 (March 2013), the central office where an OLT is located, is therefore linked to numerous ONUs over a passive splitter to the fiber to the home (FTTH). The FTTH applications, services and the required bandwidth associated with those services are summarized in Table 3.

Table 3 shows the required bandwidth per service or application to work without any hitches and the total required data per single ONU is 44 Mbps. The splitting ratio considered in this research work was 1:256 and the $C_{\max}$ used for FTTH uplink and downlink are 156 Mbps and 78 Mbps as calculated. Therefore, the integrated NG-PON-WAN with a maximum SR 1:256 different ONUs satisfied the requirements efficiently compared with 78 Mbps. The $C_{\max}$ model can be expressed as [12]:

$$C_{\max} = \frac{1}{SR} \times (R_{ul} + R_{dl}) \tag{7}$$

where SR denote as the split ratio, while the R_{ul} and R_{dl} are the uplink and downlink data rates of the integrated network. The summation of R_{ul} and R_{dl} considering an NG-PON symmetrical data rate of (10/10 Gbps) that is the total amount of data can be conveyed from an OLT to 256 different ONU within the network via an SR over a time.

4 Results and Discussion

In this work, we used Matlab and Homer Pro simulation software tools. The Matlab simulation tool was used to model the PC at an EN, CPE and the overall PC with splitting ratio up to 1:256. The hybrid optimization of multiple electric renewables

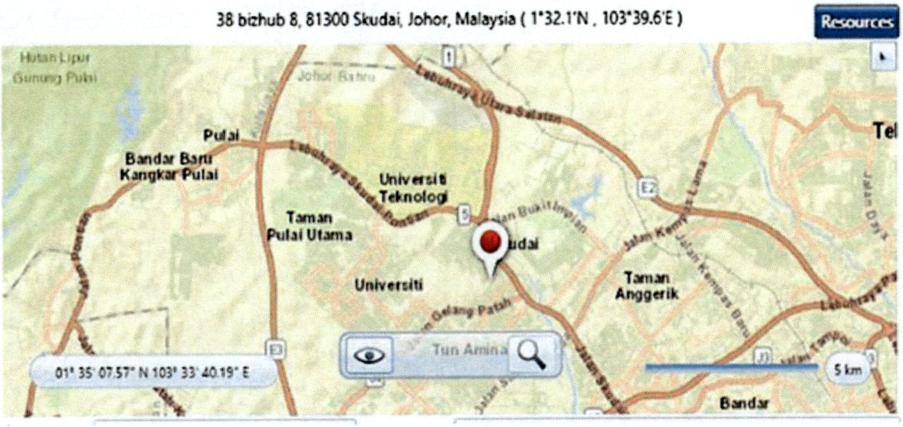

Fig. 2 Skudai-Johor, Malaysia location

(HOMER) Pro was used to estimate the designed cost for both renewable and non-renewable energy sources. The eNBs are considered the main source of PC in cellular networks which are usually located at remote areas. We consider Skudai-Johor, Malaysia as our remote location (Fig. 2).

4.1 Power Consumption Analysis

The PC of an integrated NG-PON-WAN with distribution of different split ratio (SR) which leads to a dissimilar energy consumption of the same loads was considered. This shows the variation of PC is due to the fact that the lower number of SR, the less energy consumed. The SR of 1:32 has the lower number of network devices connected to an OLT equipment through the passive device, can save power consumption of about 97.5% compared to the highest number of 1:256 SR as depicted on Fig. 3.

Figure 3 shows the simulation results for the entire nodes considered in this work. We observed the energy consumed with 1:32 SR at 2 Gbps was around 7415 W and at the same data rate with 1:256 SR, the consumption increased to around 7455 W as shown in Fig. 3. On the same figure, at 8 Gbps data rate with 32, 64, 128 and 256 different SR has the following PC of 7425, 7445, 7485 and 7565 W. Therefore, the highest SR has more PC due to number of connected devices in the network system.

At idle power and zero load, the energy consumption is as much as 90% of the total device PC and that gives 7400 W. Accordingly, from the result we analyze that considering different SR causes a linear increase in energy consumption. However, the increase in PC is caused by the daily increment for the number of

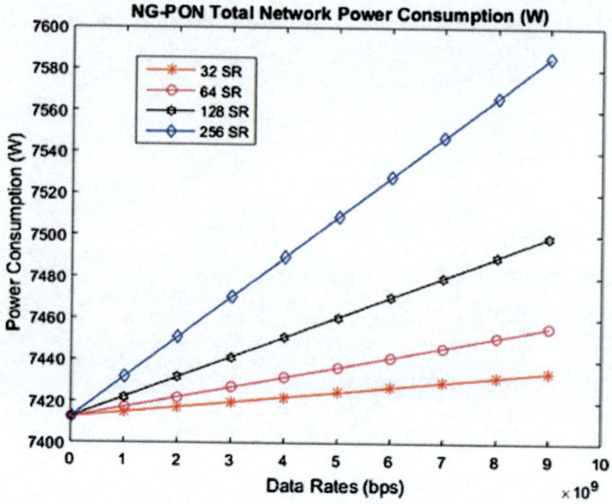

Fig. 3 NG-PON-WAN power consumption

mobile devices connected to the CPE nodes. Hence, the operational cost of the cellular companies will be increase when fossil fuel is used to power their network equipment's with a diesel generator at the remote locations.

4.2 *Renewable Energy and GHG Emission Analysis*

The renewable energy and GHG analysis have been conducted using Homer Pro software. Figure 4 shows the design diagram of the implemented for both renewables (solar-PV and wind-EO10) and non-renewable (Gen) energy resources coupled to the electrical loads (eNB1, eNB2 and eNB3) with an overall energy consumption of 186.47 kWh/day and a peak power of 15.92 kW.

In these analyses, the project life time was designed for 25 years and the load profile was generated based on the calculated PC and the sensitivity list of the components used over the period per hour. The solar PV has limitation due to sun

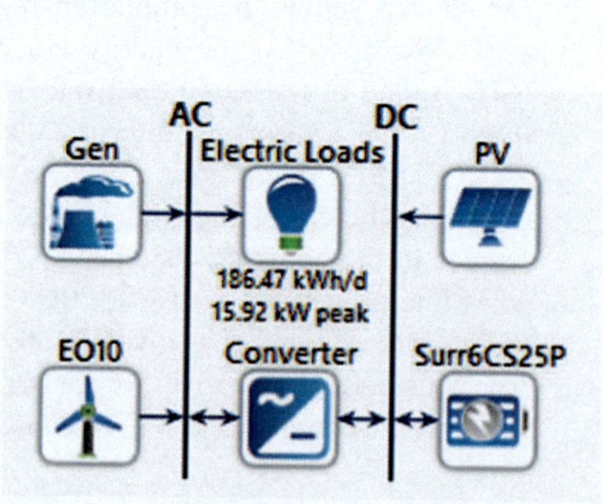

Fig. 4 Renewable and conventional energy sources schematic

availability during day time and at night the battery bank (Surr6c25P) supplies the power to the loads through converter for a period of the battery capacity. The wind (EO10) provides energy to the loads when solar PV and the battery are not accessible. The generator (Gen) system was used in the design simply to complements system's design, when other sources are under maintenance and therefore Gen can be used as a backup for maximum efficiency of the system.

The loads profile was made based on the operating hours by the CPE node. A single eNB comprises of PA, RF, BB and ONU located at an CPE node which operated 24 h per week. In this case, each eNB has some components that operated 24 h and others work for 12 h (from 18 to 6 h) only during the night when the temperature decreases and less traffic (off-peak).

Figure 5 shows the GHG emission produced per month from January to December considering the combination of solar, wind and genset architecture. The month of February (3.43%) and April (2.72%) has the less emissions compared to with July (11.27%), August (14.24%) and September (13.02%) indicating that the amount of solar radiation is higher in the first quarter compared to the third quarter of the year. The use of fossil fuel for GenSet in the third quarter of year is significant which increases the volume GHG emissions as shown in Fig. 5. However, the overall emissions released for the integrated NG-PON-WAN have been reduced by 77% within the project life time compared in [5].

4.3 Operational Cost Analysis

The system and the operational cost for an integrated NG-PON-WAN are presented on the bar chart diagram below. Figure 6 presented system cost which comprises the initial cost, operation and Maintenance (O&M) cost, replacement and fuel cost are accumulated in each component shown in Fig. 6a. The solar PV has highest cost, while the Autogenset recorded as the second highest cost. The solar PV will be the cheapest when selected for the period of 25 years of the project life time. On the

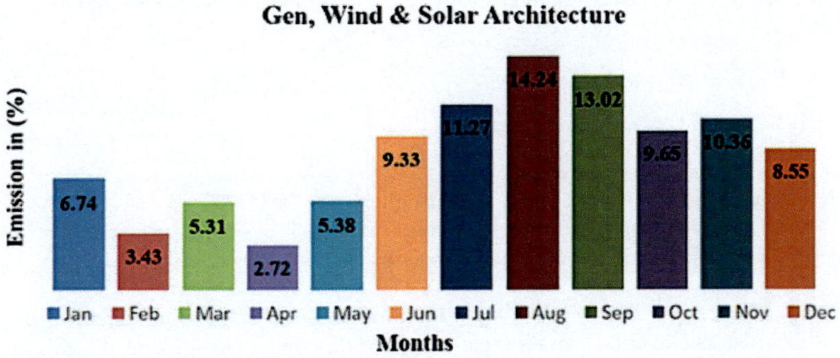

Fig. 5 Percentage of GHG emission considering solar, wind and genset architecture

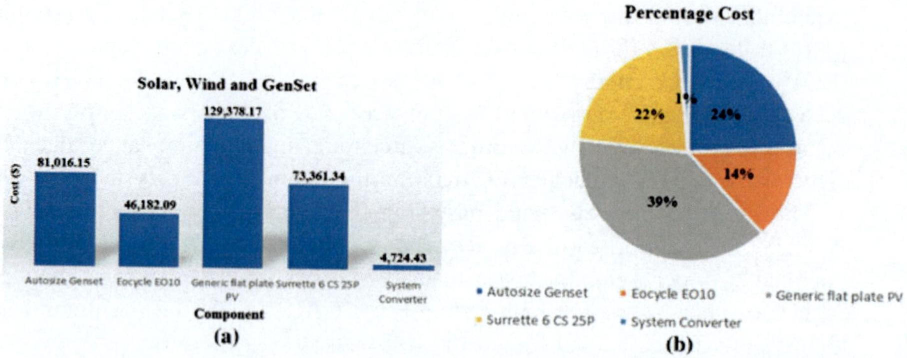

Fig. 6 Solar, wind and genset, **a** system cost and **b** percentage cost

other hand, Fig. 6b shows the percentage cost of the entire system and the highest indicated solar have 39% cost, an autogenset 24, 22% for batteries, wind turbines 14 and 1% for the system converter.

5 Conclusions

In conclusion, we presented the NG-PON-WAN system, which was modelled and simulated using Matlab software. The simulation was done in phases; the EN in the OLT location and the combined ONU-WAN at the CPE node. The Homer Pro software was used in simulating the CPE system's load with and without renewable energy sources. The renewable energy sources considered was based on the Skudai, Malaysia location and the obtained results were compared based on operational cost and CO_2 emission. A significant reduction of 77% GHG emission were achieved and 56% cost were saved for the project life time of 25 years.

Acknowledgements The authors would like to thank the Ministry of Higher Education of Malaysia and the administration of Universiti Teknologi Malaysia (UTM) for the financial assistance through Transdisciplinary Research Grant fund vote number 05G80.

References

1. Ricciardi S, Palmieri F, Fiore U, Castiglione A, Santos-Boada G (2013) Modeling energy consumption in next-generation wireless access-over-WDM networks with hybrid power sources. Math Comput Model 58(5–6):1389–1404. https://doi.org/10.1016/j.mcm.2012.12.004
2. Zhu M, Zeng X, Lin Y, Sun X (2017) Modeling and analysis of watchful sleep mode with different sleep period variation patterns in PON power management. J Opt Commun. Netw. https://doi.org/10.1364/jocn.9.000803

3. Alsharif MH, Nordin R, Ismail M (2015) Energy optimisation of hybrid off-grid system for remote telecommunication base station deployment in Malaysia. Eurasip J Wirel Commun Netw 2015(1):1–15. https://doi.org/10.1186/s13638-015-0284-7
4. Ramli A, Zulkifli N, Idrus SM (2017) Power consumption modeling and analysis of integrated optical-wireless access network. Int J Electr Comput Eng 7(6):3475–3483. https://doi.org/10.11591/ijece.v7i6
5. Aderemi BA, Daniel Chowdhury SP, Olwal TO, Abu-Mahfouz AM (2018) Techno-economic feasibility of hybrid solar photovoltaic and battery energy storage power system for a mobile cellular base station in Soshanguve, South Africa. Energies 11(6). https://doi.org/10.3390/en11061572
6. Al-Shammari HQ, Lawey A, El-Gorashi T, Elmirghani JMH (2019) Energy efficient service embedding in IoT over PON. In: International conference transparent optical networks, vol 2019-July, no April. https://doi.org/10.1109/icton.2019.8840429
7. OLT Cisco ME 4600 series optical line terminal, pp 1–8
8. Cisco (2013) Cisco ME 4600 series optical network terminal data sheet, pp 1–8
9. Kumagai Y, Funyu Y, Maeda H (2012) High-efficiency power amplifier for LTE/ W-CDMA system
10. Watanabe K, Machida M (2012) Outdoor LTE infrastructure equipment (eNodeB). Fujitsu Sci Tech J 48(1):27–32
11. Arnold O, Richter F, Fettweis G, Blume O (2010) Power consumption modeling of different base station types in heterogeneous cellular networks. In: 2010 Future Network & Mobile Summit, pp 1–8
12. Luo Y, Zhou X, Effenberger F, Yan X, Peng G, Qian Y, Ma Y (2013) Time-and wavelength-division multiplexed passive optical network (TWDM-PON) for next-generation PON stage 2 (NG-PON2). J Lightwave Technol 31(4):587–953

Integrated Circuit Characterisation Using Low-Cost Coaxial Probe Measurement

Lik Suong Ding and Xavier Ngu

Abstract This paper presents a method to characterise integrated circuit (IC) using a low cost coaxial probe. The characteristic impedance of the IC can be used in both time and frequency domain, where it significantly improves the behavioral model of a printed circuit board (PCB) circuitry which is currently being used with limited functionality. S-parameter results obtained from the measurement using the coaxial probe was combined in the simulation model of the PCB which was simulated using three-dimensional (3D) full wave simulation. A comparison between the measurement and simulation results was made and it shows that the measured IC characteristic impedance through the devised low cost probe method produces good correlation up to 1,000 MHz. This study concludes that the feasibility of using the proposed technique for impedance characterisation of IC to be included into the PCB circuitry would provide a better understanding of signal and power integrity, and electromagnetic compatibility (EMC).

Keywords Electromagnetic compatibility · Impedance · Integrated circuit · Printed circuit board · S-Parameter · Probe · Low cost

---

L. S. Ding · X. Ngu (✉)
Faculty of Electrical and Electronic Engineering, University Tun Hussein Onn Malaysia, 86400 Batu Pahat, Johor, Malaysia
e-mail: xavier@uthm.edu.my

L. S. Ding · X. Ngu
Center for Applied Electromagnetic, University Tun Hussein Onn Malaysia, 86400 Batu Pahat, Johor, Malaysia

L. S. Ding
RF Station Sdn. Bhd, A-2-7, Glomac Damansara, Jalan Damansara, 60000 Kuala Lumpur, Malaysia

© The Author(s), under exclusive license to Springer Nature Singapore Pte Ltd. 2021
Z. Zakaria and S. S. Emamian, *Advances in Electrical and Electronic Engineering and Computer Science*, Lecture Notes in Electrical Engineering 741,
https://doi.org/10.1007/978-981-33-6490-5_6

1 Introduction

Characterisation of IC is extremely important. Without good understanding of the IC, engineers are unable to correctly utilise the full potential of the IC. This is because each IC normally has typical, minimum and maximum performance. Engineers need to use the IC with correct characteristic in their simulation model to simulate the rise time and fall time along with their parasitic effect in order to maximize the value of each IC produces. Characteristic impedance is relevant to signal integrity analysis. However, Input/Output Buffer Information Specification (IBIS) models that are provided by most IC suppliers are inadequate due to the lacking of characteristic impedance information. IBIS model only provides the rise time and fall time of certain input/output (I/O) pins, but the characteristic impedance of the pins are not available. The input/output buffer information that is provided by the IBIS model is only useful for engineers to evaluate the signal integrity and the quality of the signals. The characteristic impedance is important for frequency domain analysis of the transmission line on the printed circuit board (PCB). Study done by Pucili et al. [1] has shown that IBIS has certain limitation, which includes lacking information of power supply rail stabilization, as compared to a SPICE model which is the actual or lumped components of the complete IC.

Automotive Electronics Council highlights the importance of the characterisation of the IC through a guideline [2] that recommends all automotive components suppliers to follow in order to comply to different standards worldwide. This will facilitate a standardized model for simulation in order to reduce the product development cycle, where each prototype will take months to complete and verify. When the crucial characteristic impedance information of an IC is not available, the prototype will be either over or under-designed. This wastes a lot of time, efforts and cost to the original equipment manufacturer (OEM).

Characterisation of an IC begins by obtaining the equivalent circuit of a transistor gate, by representing a transistor with an Equivalent-Tee circuits [3]. However, representing a complex IC into many Equivalent-Tee manually is a complicated task and would waste a lot of time and cost. Many researchers [4] have shown many complex measurement methods to extract data of an IC, in which correct termination and shorting the internal connections to accurately capture the resistance, capacitance and inductance of each pins are required. This method is not feasible for application engineers, who only have access to the commercialize ICs.

A recent research paper [5] summarises the existing methodologies available to characterise ICs and interconnects. IC characterisation is crucial because it improves the efficiency of the design and development stage. Proposals to use frequency domain and time domain were also introduced in the summary shown in [5].

Recently Jeong [6] performed a study on electrical characterisation of a ball grid array (BGA) using S-parameter method, and compared it with other traditional methods like the impedance measurement and the Time Domain Reflectometry (TDR) methods. The results show that by using one-port S-Parameter method the measurement time is a lot faster and the accuracy can be kept within 0.3% for frequency up to 500 MHz. However, this method requires expensive probe station and it neglects the coupling between the pins. Another study [7] on BGA was done using a microprobe station to study the coupling effect between tracks. It was found that the frequency domain method is more accurate when used with the correct de-embedding method to remove the effect of the cable losses, fixture effect and parasitic effect. The disadvantage of this method is the need to prepare a customised BGA board to measure the open, short, load, and thru. This is not feasible in many applications where limited resources are available and also engineers are normally required to characterise more than one IC. To characterise all ICs in their applications, the job is simply too overwhelming.

While it is important to characterise an IC, it is also equally important to characterise the used test board. Wang et al. [8] demonstrated a noble way to characterise all components on a low-dropout (LDO) voltage regulator test board. After characterising all the components on the test board, an accurate model of the LDO IC can be obtained up to 2,000 MHz. The non-linear impedance of the IC and test board are captured reasonably well and translated into equivalent circuits of individual RLC components. However, the completed equivalent circuits are still unable to fully correlate with the characteristic of the LDO's immunity at high frequency. This can be improved by representing the equivalent circuits with higher order equivalent circuits.

Another way of characterising the IC is by using RF micro-probes [9]. High precision RF micro-probes can be used to measure the IC impedance up to 110 GHz [10]. However, RF micro-probes are very expensive and must be used delicately to ensure correct probing of the IC. They must be handled with care and often only dedicated experts are allowed to operate the machines. Calibration techniques of the RF micro-probes are also very crucial in order to measure the characteristic of the IC correctly.

Another challenge when measuring the IC impedance is that the calibration technique requires the IC to be shorted or loaded internally. This is impossible for engineers because only the manufacturers has access to the internal circuits of the IC only. The manufacturers can customise the IC to the desired conditions for calibration purposes. Their engineers can then calibrate their probes or boards using these ICs for Open-Short-Load calibration before actually measuring the final IC impedance. This allows the engineers to obtain more accurate data up to 3 GHz [11]. However, this is not possible for user other than the manufacturers themselves where they only have access to off-the-shelf IC that is commercially available.

2 IC Impedance Measurement Using Low-Cost Coaxial Probe

2.1 Low-Cost Coaxial Probe Fabrication

For the measurement of the IC impedance in this study, a customised probe was produced. A 50 Ω RG402 semi rigid cable was used to create the probe. Semi rigid cable was used because of its stable performance and ability to connect to the 50 Ω cables with different connectors. The cable was connected to a female 3.5 mm coaxial connector on one end, A-Side, and the other end was left open, B-Side, as shown in Fig. 1. The B-Side was prepared to connect and disconnect from a male 3.5 mm coaxial connector acting as a connection to the calibration kit. When the B-side's coaxial is disconnected, it will be used as the probe to measure the chip impedance. During the calibration, the B-Side will be connected to the male 3.5 mm coaxial connector securely to ensure the ground of the semi rigid cable is connected properly. For the calibration, Rohde & Schwarz R&S®ZV-Z135 Calibration Kits was used because of its performance up to 15 GHz. This method is more accurate as compared to customised Thru-Reflect-Line (TRL) calibration board due to the broadband requirement.

In the first design of the probe, a spring loaded pogo pin is connected to the outer shell of the coaxial cable as shown in Fig. 1. The pogo pin provides a good contact during the measurement because of the spring inside the pogo pin, but it lacks flexibility in terms of the gap between the center conductor and the pogo pin. This is problematic if different sizes of ICs are required to be measured. Hence, a revised probe was designed using a copper wire, which can be bent to accommodate contact at different distances. The revised probe is shown in Fig. 2.

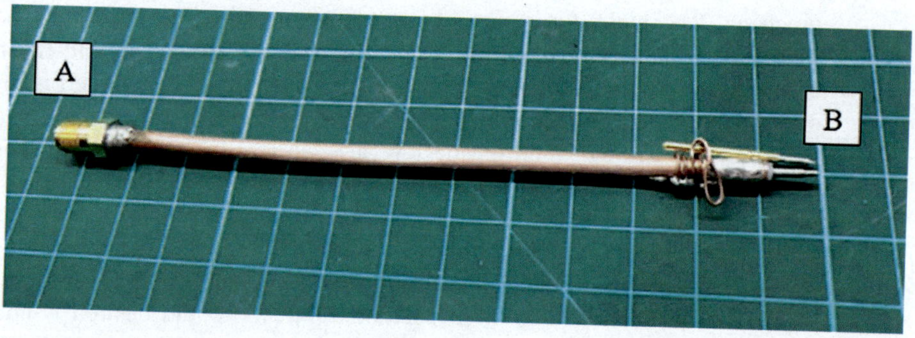

Fig. 1 Customised probe with pogo pin

Fig. 2 Revised probe with copper wire, Teflon insert and copper tape

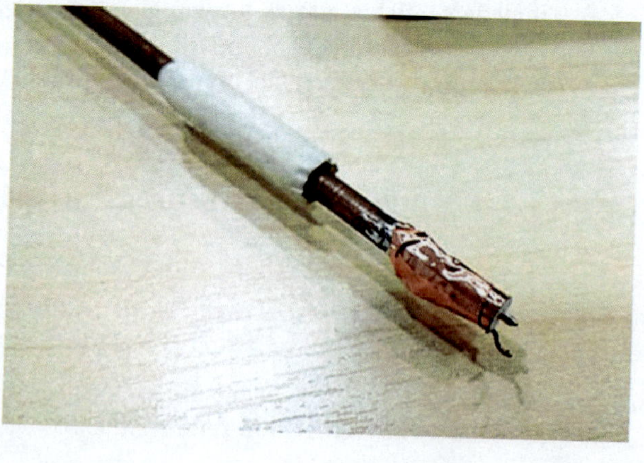

To reduce stray parasitic inductance and capacitance caused by the wire, the Teflon insert from the coax connector was also left at the tip of the probe to shorten the length of the center conductor. Copper tape is then used to wrap the wire, coaxial cable and the Teflon insert to maintain a constant 50 Ω as much as possible. This final setup is more similar to the physical setup of the coax connector when it is connected to the 3.5 mm SMA calibration kit.

2.2 *Test Board for IC Characterisation*

IC used in this study is an Electromagnetic Interference (EMI) filter at line termination for USB upstream ports from STMicroelectronics. Model of this IC is USBUF [12]. This IC was chosen because it is one of the most popular device used in modern devices due to its small footprint and its cost effectiveness. This IC not only filters the EMI from the PCB but also used to withstand Electrostatic Discharge (ESD) up to 15 kV air discharge. In this study, the challenge is to handle this device due to its size, the dimension is merely 2 mm (length) by 1.25 mm (height). Each pin's width is only 0.3 mm. The functional diagram and dimension of the IC are shown in Fig. 3.

To measure the S-Parameter of the IC, a customised FR4 PCB test board was fabricated to hold the IC (test board A) as shown in Fig. 4. The IC's ground pin and the 3.3 V voltage input pin were shorted to the PCB ground through vias with the bottom ground. Other pads are then connected to 50 Ω surface-mount device (SMD) resistors to perfectly matched probe's impedance. First, test board A was set up by leaving pin 1 and 6 open, pin 2 and 5 shorted to ground, and pin 3 and 4 soldered to 50 Ω resistors. The test board was held in place by a fixture to prevent movement during the probing measurement. The probes were held stable by a rotary arm to prevent accidental movement while capturing the data obtained from a

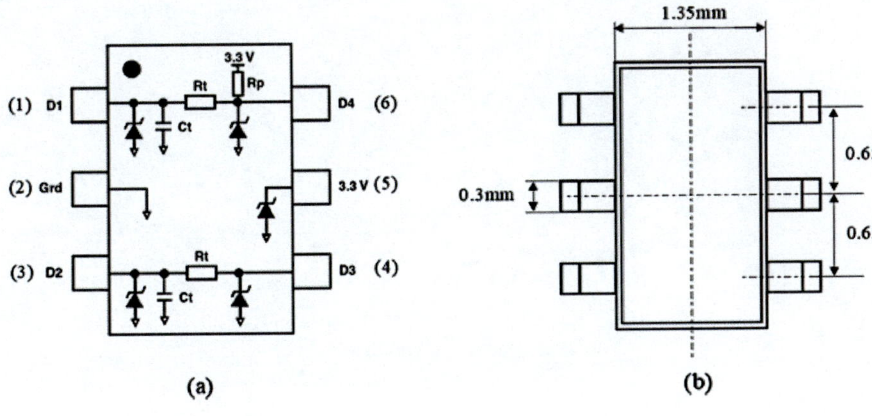

Fig. 3 **a** Functional diagram of the IC, and **b** package dimension of the IC

Vector Network Analyzer (VNA) from R&S® model ZVB8. Pin 1 and pin 6 were measured with port 1 and port 2 respectively. Ferrite cores were installed along the length of the cable to prevent emission due to the unbalanced current at the outer layer of the coaxial cable. The measurement setup is shown in Fig. 5.

After measuring pin 1 and 6, the obtained S-Parameters of the IC are stored as touchstone file, namely s2p to be used later in test board B simulation. The S-Parameter results are shown in Fig. 6. A slight resonant can be observed at around 1,700 MHz for both S1,1 and S2,2 which is due to the parasitic effect of the copper wire. When both probes are near to each other, coupling between them becomes stronger and they act as a miniature antennas. This coupling is difficult to prevent because the ground wire is needed to maintain flexibility and at the same time measures the S-Parameters of the IC at various locations.

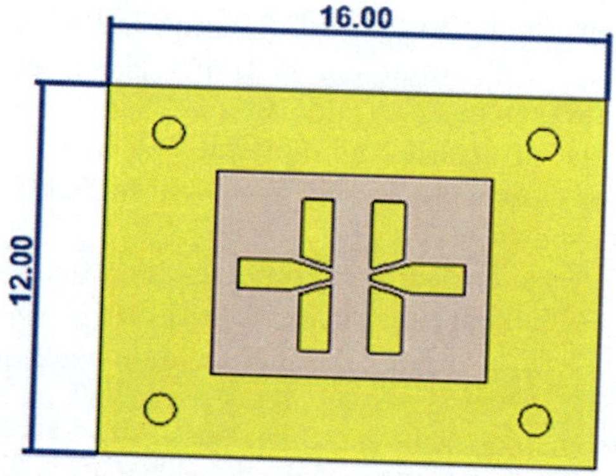

Fig. 4 Test board A to measure IC impedance

Fig. 5 Measurement setup of test board A with probes being held by a fixture

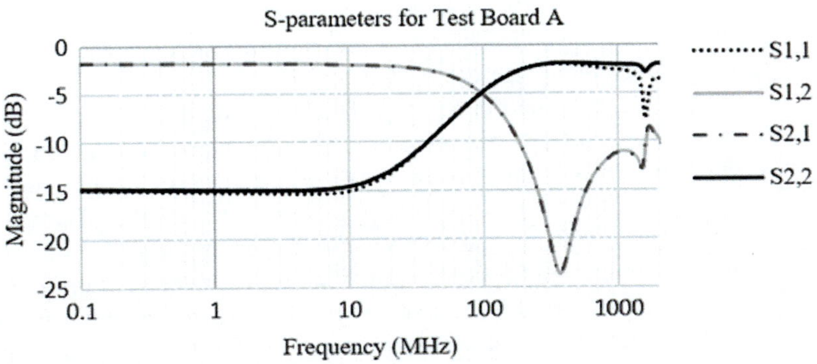

Fig. 6 Measured S-parameters on test board A

3 IC Characteristic Impedance Implementation

3.1 Combining the S-Parameters of Simulated Test Board with Measured IC Characteristic Impedance

Another test board (test board B) was fabricated to verify the IC characteristic impedance measured in test board A. Test board B as shown in Fig. 7 (a) was fabricated based on the datasheet from STMicroelectronics as shown in Fig. 7 (b). Since test board B has less discontinuity i.e. close to 50 Ω characteristic impedance transmission line and connected to a 50 Ω SMA connector, test board B shall yield better result. The aim here is to compare the results between test board A and B. To maintain a 50 Ω transmission line from the coaxial connector, coplanar waveguide with ground (CPWG) transmission line was used in test board B. In order to achieve 50 Ω on the CPWG transmission line with a 1.6 mm FR4 PCB, the track width is 2.5 mm with ground clearance of 1 mm. The width of the transmission line is very different as compared to the datasheet because the manufacturer did not mention the material used for the PCB as well as the transmission line physical properties. Hence, FR4 is used for the test board B due to this limited resources. For this test setup, the ground pin (pin 2) of the IC was shorted to the voltage input pin (pin 5), while pin 3 and pin 4 are left open, as documented in the datasheet model.

Test board B is then modeled in 3D within Computer Simulation Technology (CST) software. CST is a full wave 3D electromagnetic (EM) simulation software that can accurately simulate the behavior of the EM wave across 3D structure based

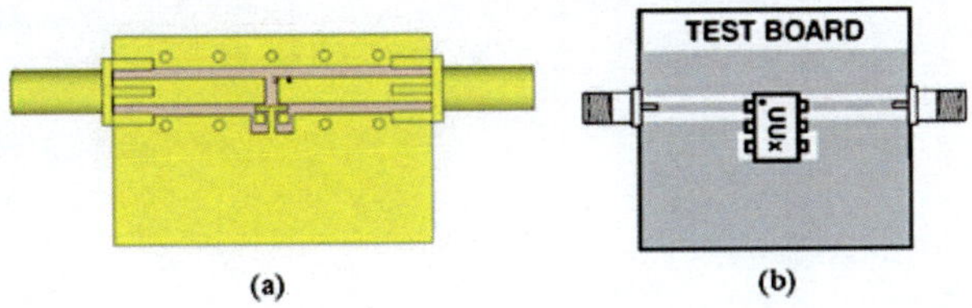

Fig. 7 **a** Fabricated test board B based on datasheet, and **b** test board model from datasheet

on Maxwell Equation [13]. In the simulation model, the SMA connectors were also included to improve the realism of the transmission line. The vias location are also similar to the actual fabricated PCB.

In the 3D simulation of test board B, ports are defined to capture the S-Parameters. Port 1 is defined on the SMA connector on the left, port 2 is defined at pin 1, port 3 is defined on the SMA connector on the right, and lastly port 4 is defined at pin 6. Simulated 4-ports S-Parameters of the 3D model can then be renormalised in Design Studio which is the circuit simulator of the CST.

Within the Design Studio as shown in Fig. 8, the block with four pins represents the simulated four-port S-Parameters. Each pin represents the simulated ports from the 3D model. Pin 1 and pin 3 are connected to the excitation Port 1 and Port 2 respectively, while pin 2 and pin 4 are connected to the measured touchstone file acquired from test board A. This touchstone file is the s2p of the IC's S-Parameters acquired from the probe. In the Design Studio, new S-Parameter results are simulated in which the 3D results are renormalised by including the IC characteristic impedance.

Test board B is also fabricated based on the exact dimension in the simulation model as shown in Fig. 9. Test board A has the practicality to be implemented in existing PCB, but the accuracy of this method must be proven by comparing results obtained from test board B which are more accurate due to the continuity of the 50 Ω characteristic impedance on the CPWG transmission line.

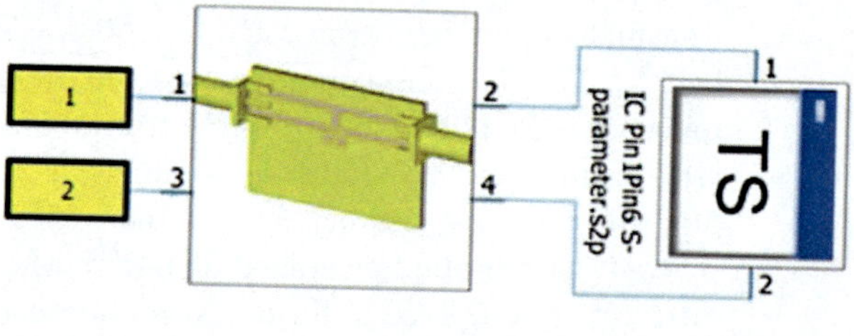

Fig. 8 Test board B 3D model configuration inside CST design studio using measured touchstone file from the measurements

Fig. 9 Test board B measurement of pin 1 and pin 6 through CPWG transmission line with SMA connectors

3.2 Characteristic Impedance Verification with Test Board

Figure 10 shows the comparison between results obtained from Test board B measurement and the simulated 3D model. Both measured and simulated results have identical S1,1 and S2,2, due to the symmetry transmission line. Hence the graph is simplified to improve visibility of the comparison. It is observed that both the return loss (S1,1) and the transmission (S1,2) correlate reasonably well up to 1,000 MHz. A resonant in S2,1 at 400 MHz can be noticed in both the results. Transmission is also good from 0.1 MHz to 100 MHz at around −2 dB to −5 dB. There is a mismatch in S2,1 between 1,000 MHz and 2,000 MHz which could be caused by the said imperfection of the copper wire from the probe. It is already known from above chapter, that there is resonant from the copper wire at high frequency, therefore only up to 1 GHz result should be focused on. The simulated results is also identical to the probe results from test board A (Fig. 6), which means that the transmission line from the test board B has minimum impact to the characteristic impedance of the IC.

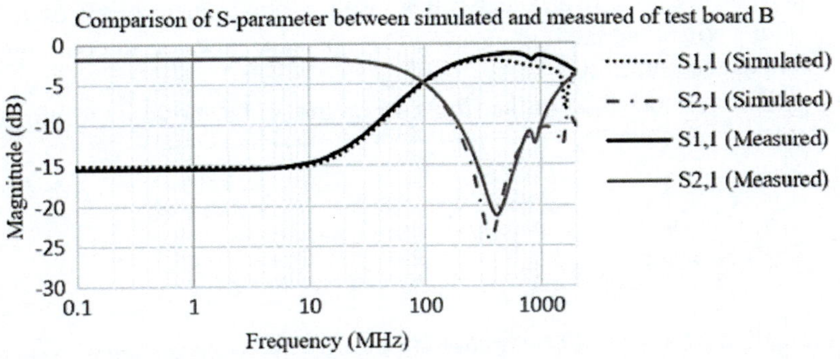

Fig. 10 Comparing S-Parameter results obtained from (simulated) with the (measured) test board B

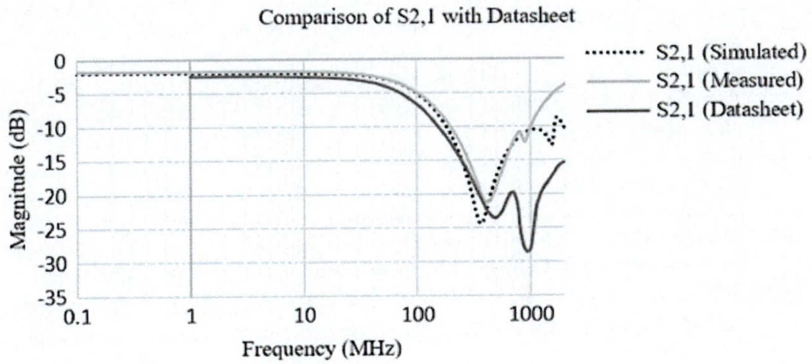

Fig. 11 Comparison of S2,1 between simulated and measured test board B, and the manufacturer's datasheet

Comparison was also made between the declared S2,1 from the datasheet of the IC manufacturer, fabricated test board B and the simulated results as shown in Fig. 11. First resonant is present on all of the results at around 500 MHz but the second resonant at 1,000 MHz is not available in both measured test board B and simulated results. This may be due to the difference on the test boards. The manufacturer does not disclose the PCB design specification and the track width of the test board, which is obviously much narrower as compared to the fabricated test board B. The effect captured in the datasheet is not the IC characteristic alone but also included the effect of tracks and pads on the PCB.

4 Conclusion and Future Work

The paper has shown that the method of using a low cost coaxial cable is able to characterise the impedance of the IC pins. This method was used on the test board A and later included the characteristic impedance of the IC in the simulation model of test board B. The results are then compared with fabricated test board B, which technically should produce better result. Both results correlates reasonably well. This verifies the method of using low cost coaxial cable in obtaining characteristic impedance of an IC.

Future work is to further improve on the probe and to perform similar approach on a more complicated 64 pins quad flat package (QFP) IC.

References

1. Pulici P et al (2008) A modified IBIS model aimed at signal integrity analysis of systems in package. IEEE Trans Circuits Syst I Regul Pap 55(7):1921–1928
2. Sullivan MO et al (2013) Guideline for characterization of
3. Pritchard RL (1998) Transistor equivalent circuits. Proc IEEE 86(1):150–162
4. Caggiano MF, Ou J, Bulumulla S, Lischner D (2001) RF electrical measurements of fine pitch BGA packages. IEEE Trans Compon Packag Technol 24(2):233–239
5. McGibney E, Barrett J (2006) An overview of electrical characterization techniques and theory for IC packages and interconnects. IEEE Trans Adv Packag 29(1):131–139
6. Jeong J (1999) Electrical characterization of ball grid array packages from S-parameter measurements below 500 MHz. IEEE Trans Adv Packag 22(3):343–347
7. Horng TS, Tseng A, Huang HH, Wu SM, Lee JJ (1998) Comparison of advanced measurement and modeling techniques for electrical characterization of ball grid array packages. Proc Electron Compon Technol Conf F1334:1464–1471
8. Wang Z, Zhou C, Member S, Liu T, Zhao S, Liang Z (2016) Nonlinear behavior immunity modeling of an LDO voltage regulator under conducted EMI, pp 1–9
9. Bullard B, McNamar R, McNally S (2018) RF probing of custom ASIC's. In: 2005 66th ARFTG microwave measurement conference: digital communication system metrics, ARFTG 2005, pp 1–7

10. Sattler SW, Gadringer ME, Gentili F, Alterkawi ABA, Bösch W (2017) PCB RF probe landing pads for multiline deembedding. In: PRIME 2017—13th conference on PhD research in microelectronics and electronics, proceedings, vol 1, pp 209–212
11. Boyer A, Sicard E, Fer M, Courau L (2008) Electrical characterization of a 64 ball grid array package. In: IEEE international symposium on electromagnetic compatibility, vol 1
12. STMicroelectronics. USBUF—2-line EMI filter and ESD protection with R pull-up, for USB interfaces. Available: https://www.st.com/en/emi-filtering-and-signal-conditioning/usbuf.html . Accessed on 26 Mar 2020
13. Vandenbosch GAE, Vasylchenko A (2012) A practical guide to 3D electromagnetic software tools. Microstrip Antennas 1:23–38

A More Efficient and Accurate Approach to Characterise an IC into Macro Model

Lik Suong Ding and Xavier Ngu

Abstract Traditional workflow to obtain macro model of an Integrated Circuit (IC) requires expensive probing measurement system as well as complex in-house program. In this paper, a more practical method to create an accurate macro model of an IC is introduced. IC's characteristic impedance has to be measured using a special probe be-cause most of the circuitry are proprietary and inaccessible to engineers. After acquiring all the combinations of the IC's characteristic impedance, the complete sets of the measured results are compiled by using a software specifically written for this purpose. The compiled results are then converted into a macro model where passivity and causality are then enforced. The macro model has to be passive to correctly represent the actual behavior of the IC. Passivity and causality of the model are enforced by using a commercially available tool and is further simulated using Computer Simulation Technology (CST) software.

Keywords Electomagnetic compatibility · Characteristic impedance · Integrated circuit · Printed circuit board · S-Parameter · Macro modeling · Passivity · Causality · SPICE

L. S. Ding · X. Ngu (✉)
Faculty of Electrical and Electronic Engineering, University Tun Hussein Onn Malaysia, 86400 Batu Pahat, Johor, Malaysia
e-mail: xavier@uthm.edu.my

L. S. Ding · X. Ngu
Center for Applied Electromagnetic, University Tun Hussein Onn Malaysia, 86400 Batu Pahat, Johor, Malaysia

L. S. Ding
RF Station Sdn. Bhd, A-2-7, Glomac Damansara, Jalan Damansara, 60000 Kuala Lumpur, Malaysia

© The Author(s), under exclusive license to Springer Nature Singapore Pte Ltd. 2021

Z. Zakaria and S. S. Emamian, *Advances in Electrical and Electronic Engineering and Computer Science*, Lecture Notes in Electrical Engineering 741,
https://doi.org/10.1007/978-981-33-6490-5_7

1 Introduction

In this paper, an Integrated Circuit (IC) called Digital Transmitter from an Advanced Driver Assistance System (ADAS) evaluation board was chosen as the Device Under Test (DUT) where the IC internal impedance are known only to the chip manufacturer. To characterise the IC correctly for all simulation purposes, a novel method is required to acquire the characteristic impedance quickly and accurately. This work will ensure that future simulations no longer assume an IC as a black box. A black box is a condition where there is no coupling between the pins internally. As a results, this simulation produces less accurate results. The workflow of this work is shown in Fig. 1.

There are many methods to characterise multi-ports device using only two (2) ports vector network analyzer (VNA), such as the work done by Tippet and Speciale [1], where a method to renormalise the measured Scattering Parameter (S-parameter) by a known impedance was introduced. In a S-Parameter measurement, each port will be terminated with a known load, which is normally 50 **Ω**. However, the mismatched load will introduce parasitic effects into the measured S-Parameter results. To counter this shortfall, Kam [2] performed a conversion with an error-correction method to renormalise the measured results with a known termination.

However, this method requires extensive understanding of the mathematics behind the conversion in order to perform this calculation accurately. Furthermore, it is only logical to perform this conversion by using a software because the ICs used in commercial are normally more than four pins.

There is also a limitation where the calculation requires a correctly determined termination [3] in order to solve the unknown in the calculation. This unknown normally comes from the tracks which are connected to the IC pins. Without accurate measurement of the termination, it will produce errors in the mathematical

Fig. 1 Project workflow

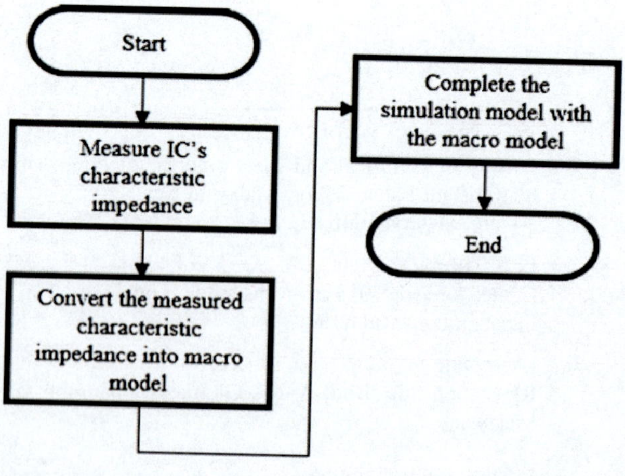

calculation matrices. Without a clear understanding of the probes, the DUT is almost impossible to be correctly characterised.

In this paper, a method is introduced by terminating the IC with 50 Ω resistor networks. In this work, the IC is characterised up to 1,000 MHz only due to the limitation of the equipments. The characteristic impedance of the IC is measured in terms of S-Parameter, which is normally used in frequency domain analysis.

After characterising the IC, the S-Parameters of the IC are converted into a macro model [4], namely the SPICE model. SPICE model is a representation of a component in terms of equivalent circuits, which consist of resistor-inductor-capacitor (RLC) lumped components. A SPICE circuit is more practical in real cases because it can be used in both time and frequency domains. On the other hand, a touchstone file which contains only the frequency response data has two noticeable limitations. Firstly, the data beyond the measured frequency will not be known e.g. frequency response at 2,000 MHz could not be obtained or estimated from a S-Parameter result of 0–1,000 MHz. Secondly, to obtain the time domain response, vector fitting has to be performed in each of the simulations to estimate the time domain response based on the frequency data. This increases the simulation time significantly. In addition, there is also a limitation due to the vector fitting algorithm used in the simulation software.

The SPICE model on the other hand is generated based on a certain level of order of vector fittings by using RLC lumped components in order to synthesise the frequency domain responses [5]. There is no need to perform further vector fitting in the simulations because the accuracy is reasonably acceptable from direct current (DC) up to the provided frequency range. Higher frequency results can also be simulated and measured, but it is necessary to maintain the accuracy of the SPICE model because at higher frequency, higher order resonants will also be present. Without characterising these higher order resonants in the SPICE model, the model beyond its frequency range should not be used.

After converting the characteristic impedance of the IC into a SPICE model, the causality and passivity enforcements will be performed on the model. This is because most of the measurement will contain certain level of parasitic effects, for example a small resonant or a spike where it is normally difficult to notice. RLC lumped components are always passive, where the energy output will never be higher than the energy input. A causal system means that the response of an output signal will never be earlier than the input signal [6, 7, 8]. Both causality and passivity enforcements will further improve the accuracy of the SPICE model which is normally introduced by human error and/or measurement devices.

2 IC Characterisation Using a Two-Port VNA

To successfully characterise the IC, measurements on all relevant pins are performed using a two-port vector network analyser (VNA), to obtain the S-Parameters of the pins i.e. S1,1, S1,2, S2,1 and S2,2. In this work, Texas Instrument's

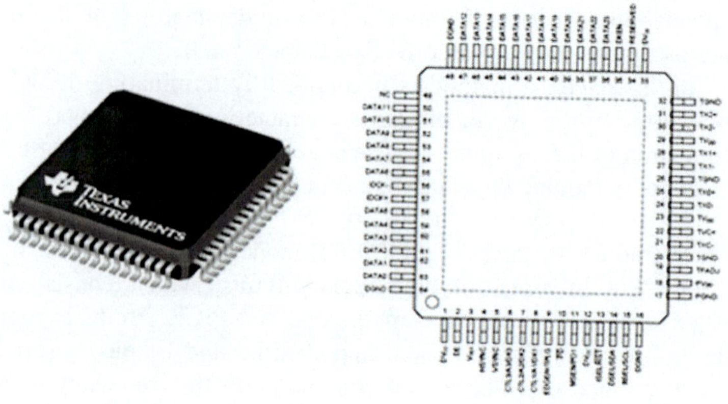

(a) (b)

Fig. 2 **a** TI PanelBus digital transmitter and **b** pins configuration [9]

TFP410 TI PanelBus™ Digital Transmitter was chosen because it is commonly used in many ADAS system to interface with display panel such as the HDMI interface. This IC is housed in a 64 pins Thin Quad Flat Package (TQFP) with different combinations of input/output (I/O) pins, differential pins, as well as power and ground pins. The physical structure and pin configuration are shown in Fig. 2.

In the S-Parameter measurement, all the pins except power and ground pins will be characterised. The reason of the exclusion is because both the power and ground pins are designed to have low internal resistance and they are connected directly to the power planes and ground planes. The power and ground pins need to have low impedance in order to ensure smooth power delivery, which will not be explained in details in this work. During the characterisation, the power pins and ground pins are shorted to ground. Out of the 64 pins, only 48 pins are required to be characterised, because the remaining pins are power and ground pins.

Prior to the measurement, the IC is soldered onto a breakout board. The power pins and ground pins are shorted to a 150 mm by 150 mm ground plane, while the remaining 48 pins are terminated with 50 Ω series resistors networks. The breakout board is shown in Fig. 3, where the breakout board is slightly elevated with a 3D printed support, to prevent the exposed soldered connections of the series resistors from shorting to the ground plane.

In another work, a low cost coaxial probe was designed using a pogo pin as a solid connection to the ground. The pogo pin is soldered to the ground of the coaxial to restrict the pogo pins from movement during the probe measurement. The probe is shown in Fig. 4. The coaxial probe is connected to the VNA so that S-Parameter measurement can be performed on the IC pins.

The measurement flowchart is shown in Fig. 5. All possible combination of the pin pairs are selected by using the combination equation in (1) which is also known as "n choose r" where it shows the number of ways to choose from r elements from a total number of n objects. The equation states that for a 48 pins configuration,

Fig. 3 The IC soldered onto a breakout board with series resistor networks for termination

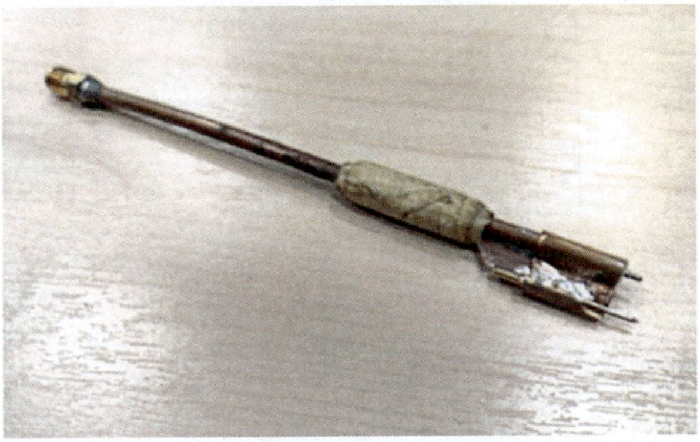

Fig. 4 A low cost coaxial probe with a pogo pin soldered to the ground

without repeating the same combination, there will be a total of 1,128 combinations, e.g. *S1,2*, *S1,3*, *S1,4*, …, *S47,48*, etc.

$$\frac{N}{2} = \frac{N!}{2!(N-2)!} = \frac{N \cdot (N-1)}{2} \tag{1}$$

To ease the probing process, additional grounds are connected at the corners and also at the top of the IC as shown in Fig. 6. This is to ensure that the common grounding is accessible to the probe during the measurement due to the limited distance between the inner conductor of the probe and the pogo pins (ground). The final breakout board with the probe is shown in Fig. 6 where it can be seen that the probe has to be precisely probing the intended track without touching the neighbouring track.

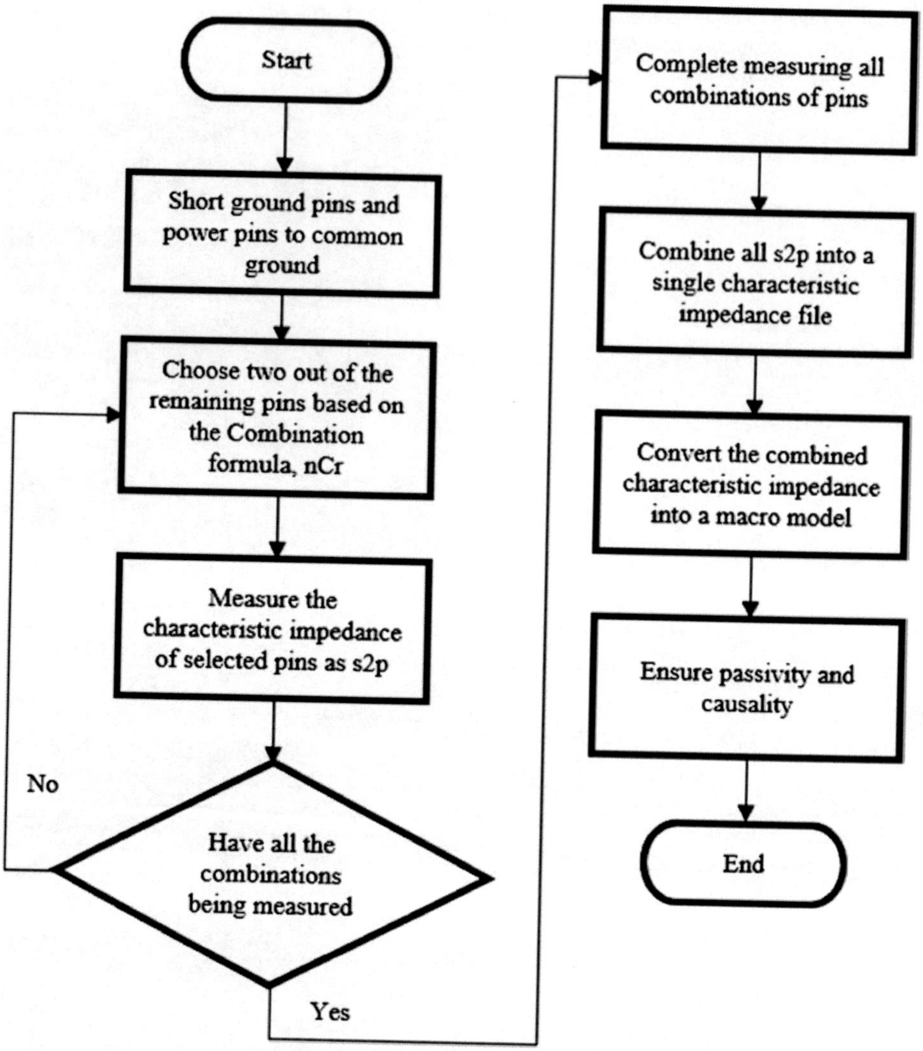

Fig. 5 The measurement process flowchart

During the measurement, the tracks have to be sliced to create an open circuit. Otherwise, they will be terminated with the 50 Ω resistors. For example, in order to obtain *S1,2*, pin 1 and pin 2 have to be disconnected from the resistor during the measurement. Next, to obtain *S1,3*, the track on pin 2 has to be reconnected by soldering it with thin wire, while the track on pin 3 has to be sliced. Due to the enormous effort and the possibility of damaging the tracks permanently if all 48 tracks have to be sliced and reconnected through the 1,128 combinations, only 10 pins are measured as to prove this concept. The pins chosen are based on their different functions, i.e. general I/O pins, differential pins and clock pins. There are

Fig. 6 Final breakout board with probe (arrows show additional ground plane)

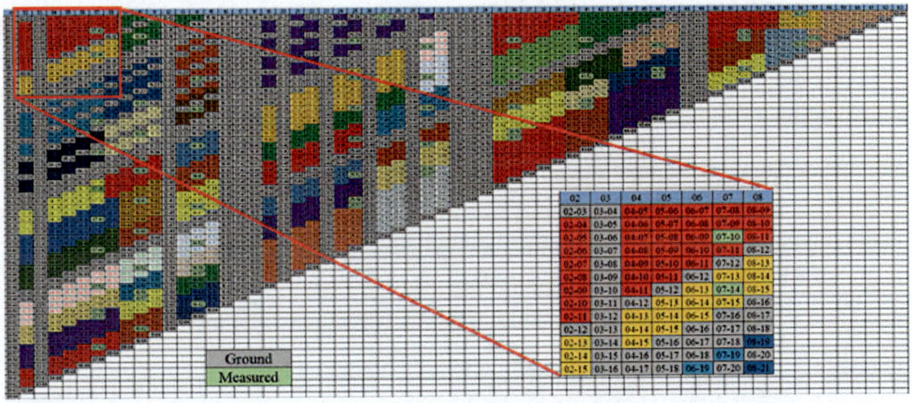

Fig. 7 Final measured pin combinations, and colour coded according to the neighbouring pins. Zoom in view for better visualisation for pin 54 to pin 63

also selected based on their location at different quadrants of the IC, in order to obtain variation information on all four quadrants. By selecting only 10 pins, the total number of measurement reduced to 55 combinations. The final measured data is shown in Fig. 7. Grey columns represent ground, while light green columns represent the measured combinations. Assuming that the neighbouring pairs will have similar coupling responses, the measured pin combinations are then applied to these neighbouring pins. They are colour coded to groups based on the measured data. Referring to Fig. 7's zoom in view, the top row represents the pin number, while 07–10 means the pair combination of pin 7 and 10. The light green color of

07–10 means that it is measured, while the red color cells around it are duplicated based on the measured results from 07 to 10. This is repeated for all combinations.

The measured data is captured in touchstone format, namely the s2p format which is the frequency response of a two-port S-Parameters. They are *S1,1*, *S1,2*, *S2,1* and *S2,2*. The combinations of s2p are replicated on pins that are not measured based on the color coded groups shown in Fig. 7. After acquiring all the s2p of 1,128 combinations out of the 48 pins, they are combined into s48p. These measured data are then combined using a self-written software where all the relevant data will be restructured from a 4 × 4 matrix to $N^2 \times N^2$ matrix, where N is the number of pins. Each s2p contains the real and imaginary part of each S-parameter i.e. *S1,1*, *S1,2*, *S2,1* and *S2,2*, therefore giving us the 4 × 4 matrix. Therefore, the s48p is a $48^2 \times 48^2$ matrix.

To perform the combination, it is more practical to use a self-written software. The software was written in JAVA due to the simplicity of utilizing the built in libraries e.g. array list library, file chooser library, file writing library and etc. After combining all the 1,128 combinations of the results, a final touchstone file is produced in a single output file which represents the s48p. Part of the touchstone file is shown in Fig. 8. A touchstone file is defined to start with the frequency point, following by real and imaginary of each data sample separated by a common space.

The s48p touchstone file is then validated in Computer Simulation Technology (CST) circuit simulator to ensure the usability. The s48p touchstone file is successfully simulated with 48 ports, and the S-Parameters of a 48-port system is obtained. Part of the simulated S-Parameters is shown in Fig. 9 where it can be noticed that the return loss (*S1,1*) is near to 0 dB at low frequency and starts to couple to neighbouring tracks at high frequency. This confirms the capacitive behaviour of an IC where during DC the signal is unable to couple to neighbouring tracks within the IC itself. The different coupling behaviour between different pins also confirms that one could never assume that the IC as a black box because at different frequencies, the coupling for *SN,1* (with $N = 2, \ldots, 48$) exists in different magnitudes between the neighbouring tracks, as well as the opposite tracks.

Fig. 8 s48p touchstone file

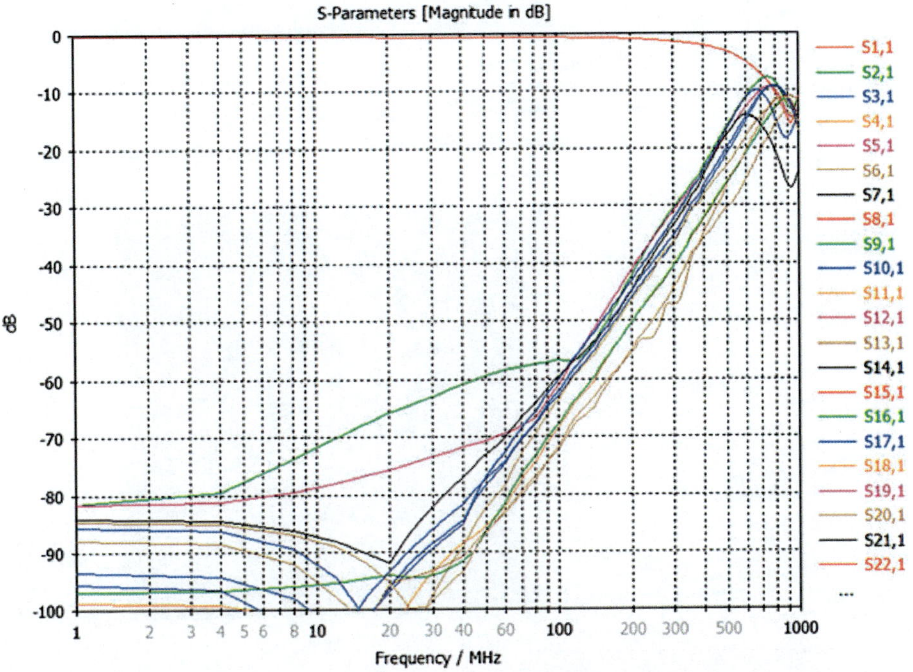

Fig. 9 Simulated sNp touchstone output file, for N = 1, 2, 3, …, 48. Showing only SN,1 where S1,1 is the return loss and SN,1 is the coupling with other tracks

It is also interesting to note that for differential pair tracks, the coupling behaves similarly to the non-differential pair tracks. Although the IC is packaged in a very small dimension, the coupling for *SN,1* (with *N* = 3,…, 48) falls below—80 dB even though they are just two to three tracks away. Similar behaviours applies to the opposite tracks. Only the neighbouring track *(S2,1)* shows higher coupling.

3 Macro Modeling Using IdEM

Finally, the s48p touchstone file is imported into macro modelling tool, IdEM [10] to generate a SPICE model, which is shown in Fig. 10. IdEM utilises state of the art vector fitting algorithm to convert the frequency response data into RLC lumped components equivalent circuits. The color within the IdEM panel is representing the dataset, where yellow color has the highest error. The error can be causality error or passivity error. Each cube in the dataset represents the behaviour of *SN,N* (where *N* = 1,…, 48)

Within IdEM, different level of orders can be used to perform the macro modeling using vector fitting. Higher order would allow vector fitting of higher accuracy, where small resonant within the results can be included in the macro model as

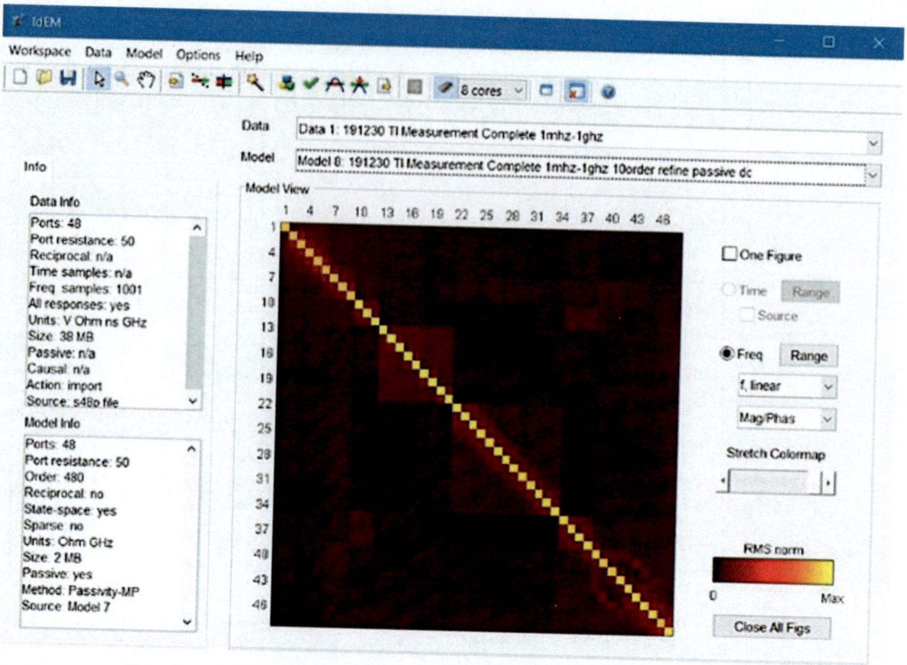

Fig. 10 IdEM macro modelling tool interface

well. But in this work, 10 orders lumped components is selected for the measured results. There is no major difference in choosing higher order except the increase of the simulation time. Hence, it is recommended to choose an acceptable accuracy within the capability of the available workstation. The targeted accuracy for this vector fitting is 10 e^{-3} and the highest error achieved is less than 0.1 in linear form. However, the accuracy of the macro model is reasonably acceptable as there is also measurement uncertainties introduced during the measurement using the probe.

From the touchstone file, the lowest measured frequency is 9 kHz due to the limitation of the VNA. This can be overcome in IdEM, where the DC point can be included during the creation of macro model, which will ensure the usability of the SPICE model for DC simulation.

The next step is to ensure passivity and causality of the macro model which is performed by utilizing Hamiltonian solver [6]. The passivity enforcement can be performed from DC to infinite frequency, but in this work it is only performed up to 1,000 MHz. The Hamiltonian algorithm will eliminate the frequency band where violations occur based on the eigenspectrum matrices. The output results are shown in singular values of the eigenvalues where more than one represent non-passive behaviour. After performing the passivity enforcement, causality enforcement is also performed. Causality enforcement is based on generalised Hilbert transform algorithm [11], where it minimises the numerical error and also it removes the truncation error which also caused the system to be non-passive.

After the passivity and causality enforcement, the output file is exported as a SPICE model with a common ground. This common ground is used as a ground reference for all the pins. This is similar to the actual IC where the pins are referenced to the ground pins available throughout the IC package. The SPICE model was then simulated in CST circuit simulator to obtain the final S-Parameters. It is shown that the coupling behaviour is now different from the measured results (shown in Fig. 9) where the coupling for *SN,1* (with $N = 2, \ldots, 48$) are spread from −48 to −85 dB, as shown in Fig. 11. In the first SPICE model, the passivity and causality checks show that the SPICE model is not passive. This could be caused by the measurement error from soldering and de-soldering of the tracks. Passivity and causality was then enforced on the SPICE model. The result is shown in Fig. 11. This means that the 48-pins equivalent circuits in the SPICE model is now causal and passive.

The time responses are also shown in Fig. 12. In this simulation, port 1 is excited with a 1,000 MHz Gaussian pulse while all other ports of port *N* ($N = 2, \ldots, 48$) are terminated with 50 Ω load. In Fig. 12 (a), touchstone file is used while in Fig. 12 (b), the converted SPICE model is used. Both results show similar responses despite some differences in coupling behaviour seen from the S-Parameters. The SPICE model can be used in both frequency and time domains and can be simulated with relatively faster speed. The time response from the

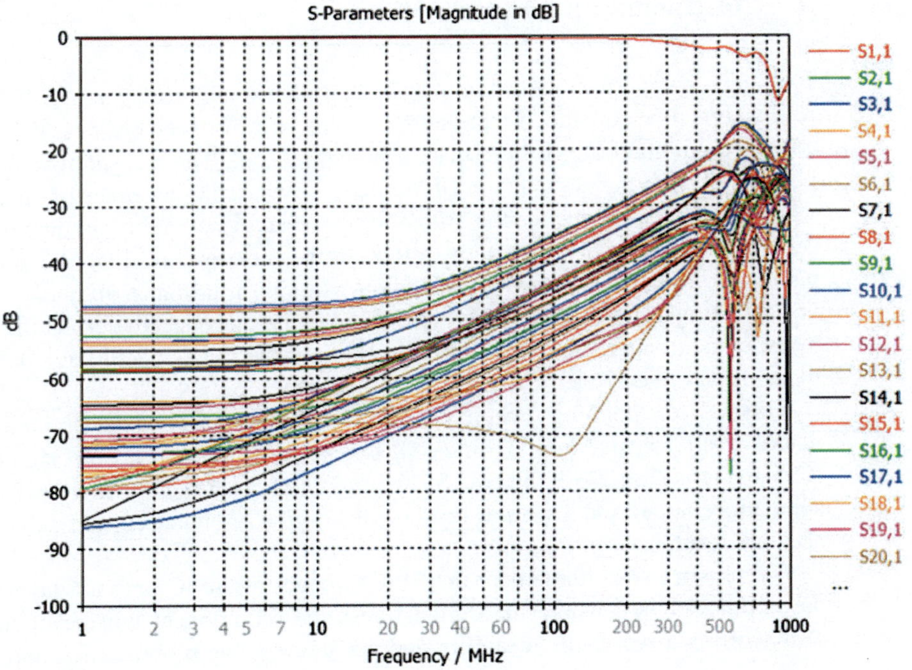

Fig. 11 Simulated *SN,1* of the SPICE model, for $N = 1, \ldots, 48$

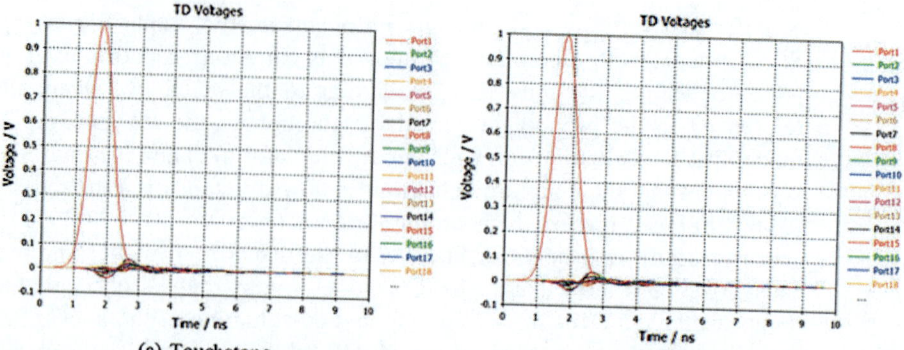

(a) Touchstone (b) SPICE

Fig. 12 Time responses of the **a** touchstone file and **b** SPICE model, where Port 1 is excited with a 1,000 MHz Gaussian pulse and Port *N* (*N* = 2, …, 48) are terminated with 50 **Ω** load

SPICE model also shows that the causality and passivity requirements are met. If a model is not passive, the output signal will have ringing behaviour after the first ripple where the ringing increases over time. In worst case scenario, the ringing will continue infinitely.

4 Conclusion and Future Work

From the comparison between measured touchstone results and SPICE model results, it can be concluded that the measurement introduced errors which was then corrected automatically by the macro modelling tool. The measurement errors can be noticed from the comparison of the dataset, where higher order of vector fitting would not be able to reduce the difference between measured S-Parameters with the equivalent circuit's S-Parameters. Although the measured results does not contain passivity violation, the passivity violation will appear at higher frequency after the vector fitting. The measurement errors could be introduced during the probing measurement, where soldering and de-soldering the tracks could have damage the transmission lines.

This new methodology where combination of low cost coaxial probe as well as market available macro modelling tool is revolutionary as it greatly simplifies the IC characterisation. With this, the SPICE model of the IC can be included in the 3D simulation model and the final result will be greatly enhanced and achieve more accurate results.

Future work is to implement the SPICE model of the IC onto a 6 layers ADAS development board where the digital transmitter component will be mounted on. This work is to verify the usability and accuracy of the proposed methodology.

References

1. Tippet JC, Speciale RA (1985) A rigorous technique for measuring the scattering matrix of a multiport device with a two-port network analyzer (comments). IEEE Trans Microw Theory Tech 33(3):286–287
2. Kam DG, Kim J (2007) Multiport measurement method using a two-port network analyzer with remaining ports unterminated. IEEE Microw Wirel Compon Lett 17(9):694–696
3. Rolfes I, Schick B (2005) Multiport method for the measurement of the scattering parameters of N-ports. IEEE Trans Microw Theory Tech 53(6 II):1990–1996
4. Schutt-Ainé JE et al (2011) Comparative study of convolution and order reduction techniques for blackbox macromodeling using scattering parameters. IEEE Trans Compon Packag Manuf Technol 1(10):1642–1650
5. Deschrijver D, Gustavsen B, Dhaene T (2008) Computer code for fast macromodeling of large multiport systems. Electr Perform Electron Packaging, EPEP, pp 299–302
6. Grivet-Talocia S, Ubolli A (2007) Passivity enforcement with relative error control. IEEE Trans Microw Theory Tech 55(11):2374–2382
7. Gustavsen B, Semlyen A (2008) Fast passivity assessment for S-parameter rational models via a half-size test matrix. IEEE Trans Microw Theory Tech 56(12):2701–2708
8. Odabasioglu A, Celik M, Pileggi LT (1998) PRIMA: Passive reduced-order interconnect macromodeling algorithm. IEEE Trans Comput Aided Des Integr Circuits Syst 17(8): 645–654
9. TFP410 TI PanelBus™ Digital Transmitter (2001)
10. Grivet-Talocia S, Ubolli A (2006) On the generation of large passive macromodels for complex interconnect structures. IEEE Trans Adv Packag 29(1):39–54
11. Triverio P, Grivet-Talocia S (2008) Robust causality characterization via generalized dispersion relations. IEEE Trans Adv Packag 31(3):579–593

Contactless Attendance Method with Face Recognition, Body Temperature Measurement and GPS System Using Blockchain Technology

Afiqah Mohammad Azahari, Arniyati Ahmad, and Syarifah Bahiyah Rahayu

Abstract Most workplace is recording their employee's attendance via attendance system. The existing attendance systems include traditional method, punch card, QR Code, RFID and biometric. However, these methods require an employee's physical interaction directly with the system. Due to COVID-19 pandemic, any direct physical contact with the system should be avoided. Moreover, the existing attendance system platforms lack the security foundations upon which vulnerable to cyber threat, such as forge by irresponsible employees and unauthorized users. This paper aims to develop a contactless attendance system. The system is using face recognition and GPS location to record attendance with body temperature on blockchain technology.

Keywords Contactless · Attendance system · Face recognition · Location detection · Blockchain

1 Introduction

Workplace attendance is fundamental importance to economy of the industry. Primarily, this appears most directly to affect those in industry as a cost of production, yet lost productivity also affects the whole economy and society directly and indirectly.

There are various methods to record employee attendance. Most organizations are recording the employee using an attendance system. The attendance system, usually, are recording employee timestamp for attending and leaving workplace.

When COVID-19 pandemic outburst, the business environment has changed to the extent that health and well-being of employees comes first. Nevertheless, employee attendance at work, their productivity and performance also become

A. Mohammad Azahari (✉) · A. Ahmad · S. B. Rahayu
Cyber Security Centre, National Defense University of Malaysia,
57000 Kuala Lumpur, Malaysia
e-mail: afiqah.azahari@upnm.edu.my

© The Author(s), under exclusive license to Springer Nature Singapore Pte Ltd. 2021
Z. Zakaria and S. S. Emamian, *Advances in Electrical and Electronic Engineering and Computer Science*, Lecture Notes in Electrical Engineering 741,
https://doi.org/10.1007/978-981-33-6490-5_8

critical component that obviously affect the performance of the business. COVID-19 pandemic changes the method of recording attendance of the employee to the workplace. Attendance system is the first point of contact for employee. The future of work and office hygiene would require the termination of physical direct contact with attendance system. Employees would be scared and unwilling to touch the biometric machines due un-hygiene spot that would spread virus. Biometric attendance system such as fingerprint method heighten security and add convenience, but this method is now being looked at as an epicenter for germs in the workplace that could include the coronavirus. This has become a potential health risk to the employees.

A new approach of contactless attendance system is expected to overcome issue of current attendance method. A contactless attendance system provides interaction between human and system without physical or direct contact with the device. The new attendance system will be able to record employee's attendance using contactless method with secured data recording. Further, this will help employers to implement fast and practical contactless attendance recording and management system, thus helping preserve a safe society.

Hence, the paper proposes contactless attendance is using face recognition algorithm, body temperature measurement and GPS location system on blockchain technology. The proposed approach is expected to overcome current issues (i.e. health and cybersecurity) and preserve a healthy workspace on top of secured technology.

2 Related Works

There are multiple methods developed for attendance recording and management system. Suitable attendance recording and management system has become important, as attendance is one of the work ethics valued by most employers. In open literature, there are multiple methods implemented using information technology for recording and managing attendance. Some methods introduced are using QR Code for scanning attendance information [1–3]. However, this method is easily manipulated by irresponsible employees. After they take a picture of displayed QR Code for scanning, they might share it with the absences. Then, the absences could scan back the QR Code and the system will record the data of absences to the attendance system. To overcome this issue, there are also methods introduce of using QR Code with Global Positioning System (GPS) [4, 5]. GPS used to authenticate location of employees when they are scanning the QR Code. However, most researchers that introduced this method did not discussed about the issue of cyber security. Not using GPS, Rahni et al. [6] introduce method of collecting students' attendances information in classes. By generating QR Code on the student's device, lecturers will scan the generated QR code from each student. This method is cumbersome, time consuming and requires lecturer's intervention. On the other hand, there are also biometrics with artificial intelligence method introduce for recording and managing attendance system. This includes fingerprint scanner and face recognition technology.

The biometric system used in attendance management system has encouraged adherence to workplace rules, all at once eliminates employee theft, 'buddy punching' and other problem and loophole caused by outdated attendance system [7–9]. However, in this current situation, method of recording attendance by using fingerprint scanner with RFID is not applicable because the devices are unsanitary and a potential source of germs, which could cause illness. Furthermore, as the novel coronavirus that causes COVID-19 could survive for 16 h on the surface, there are needs of contactless attendance system. Although, there are also contactless biometric system introduce by researchers, which is using face recognition. This method allows employees to scan their face to the special camera installed at the premises or by using CCTV and scanning the employee's face "in the wild", meaning the person does not have to pose to the scanner [10–13]. Although this method could solve the needs of contactless attendance issue, however this method required large data storage and high processing units as it need to store, recognize, and train the data of faces. Furthermore, all face recognition methods used for attendance recording and management system does not elaborate on the implementation of security for securing their stored data. Another popular biometric method is using fingerprint scanner to authenticate, identity the population and record attendance [14–16]. However, with the emerging of novel virus at the end of year 2019, fingerprint scanner as attendance recording system is impractical as the virus could transmitted via human touch. To researcher knowledge, RamcoGEEK is the only attendance system embedded with temperature recording. The system is using biometric method and thermal scanner to records employees' attendance and temperature. However, the system is still in the software testing stage.

According to Marsico et al., mobile security protocol must be heightened to preserve data confidentiality, integrity, and availability from unauthorized access. Therefore, for enhancing data security, blockchain technology will be introduced. Blockchain will be applied for securing record of employee's attendance data. Data stored in the blockchain will be transparent and immutability. Known as decentralized and distributed ledger, data stored in the blockchain cannot be modified at ease and tempered with. This is because the stored attendance data will be in a form of blocks and all the blocks are connected to each other ultimately forming a chain of records. Any changes on the block will notify other blocks for approval. This technology will prevent modification of data from third party members or from the organization itself [17].

3 Proposed Main System Components

This study is proposing a contactless attendance recording and management system. This system allows employee to record their attendance and temperature at the same time. Employees will use the system via mobile device. There are four main components of the proposed contactless attendance system as follows:

3.1 Face Recognition

Four important processes involved to recognize the employee's face. First the image for each employee will be captured and converted into greyscale. The image then stored with the ID in the database. During face detection process, a face detection algorithm will be used to identify local features of the face by dividing the image into pixels. While during feature extraction process, the Region of Interest (ROI) are parts of the face from where features are extracted and information about the gradients in the face is captured. After the completion of detecting and processing the face, the face present will be compared with data in the employee's database. When a face is detected by the camera it checks the corresponding values of the current visible face with values stored in the database. If the values are a match, then the face is recognized.

3.2 GPS Locator

There are two major process of location locator and attendance registration. The location locator will use GPS coordinates to instantly determine the employee's workplace location, based on the coordinates received. The GPS will obtain both x- and y-coordinates using ground and space satellites. The employee workplace location based on coordinates will be saved into the employee's database. The attendance system will be developed to automatically updates attendance in the database for any faces that recognized and registered. Including the x-coordinate, y-coordinate that matched the employee workplace, the employee's data and timestamp.

3.3 Temperature Measurement

The system will use a digital thermometer application from the employee's smartphone to measure their body temperature. Employee temperature will be recorded and check to ensure that the employee's body temperature is normal and able to work. Daily temperature data will be saved into the database. If the body temperature is over 37.5 °C, a notification will be sent to the organization's health department or human resource department for further action.

3.4 Blockchain Implementation

Blockchain in the attendance system will store employee records in form of blocks and validate employee certificates. Each record is treated as a block in the blockchain. The block structure consists of block header and employee data. The block header consists of hash code of the previous block, current hash value, time stamp and nonce. Each employee record is treated as a block in the blockchain. Once the block is created, a block address is created. The block is sent to the network for storage. Once the block is stored successfully, data owner will receive an acknowledgement. The certificate can be validated anytime in a future.

4 System Architecture

This contactless attendance system will introduce a new contactless attendance recording and management system with block chain technology for securing attendance data. A new mobile application will be developed, functioning for face scanning, measuring body temperature, and locating location of the employee. This research will integrate existing face detection algorithm, body temperature measurement and GPS method.

Figure 1 shows a flowchart for employee registration to the contactless attendance system. During registration, employee must bring their mobile device to be

Fig. 1 Registration flowchart

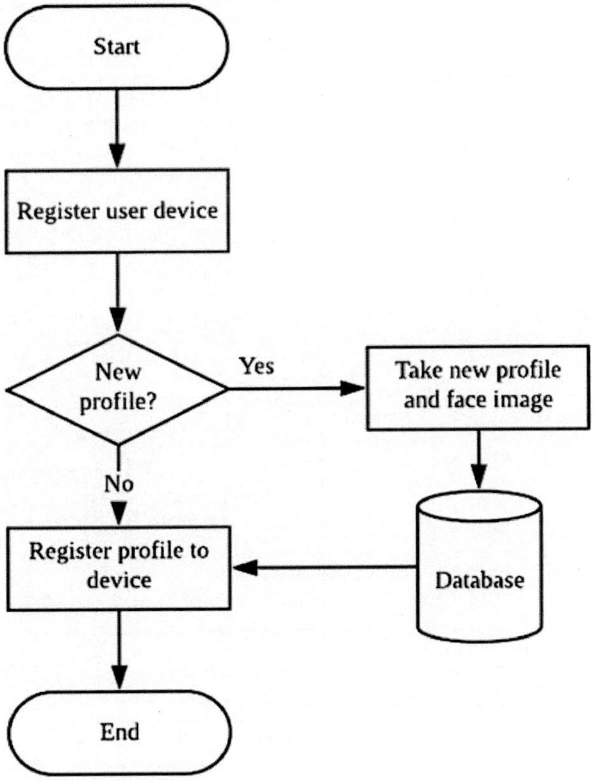

registered. A new profile will be created. Administrator will take picture of employee face. Then, the new profile and face image will be stored into a database. Then, the system will register employee profile on the mobile device. If the employee already has stored his/her profile and face image, then the system will only register employee profile to a mobile device.

Figure 2 shows the flowchart for recording employee's attendance. Employee will login to the contactless attendance system using face recognition. If it is valid, then the GPS location will be located. The GPS will be compared with the coordinates of workplace. After the GPS location and face recognition are validated, the contactless attendance system will measure employee body temperature. If the reading body temperature is above 37.5 °C, the application will send a notification to Health Department/Human Resource Department for further action. Next, information such as employee identification, clock-in, date and location of employees will be sent and stored in the distributed database. The same process will be run when employee's clock-out.

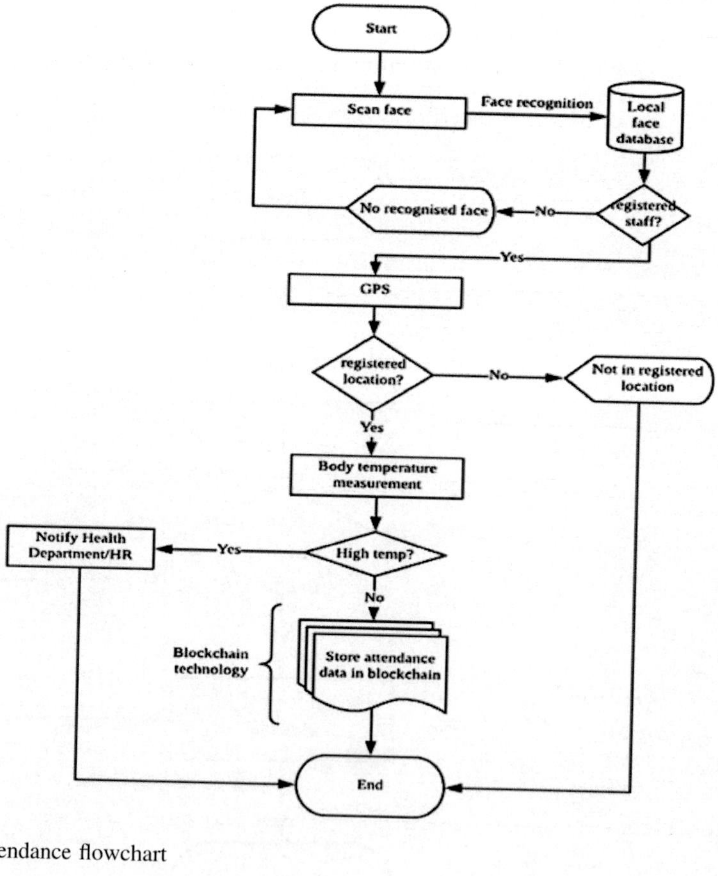

Fig. 2 Attendance flowchart

5 Conclusions

This contactless attendance system is secured and applicable for all sectors such as educations, industries, governments, tourisms etc. to record of employee attending and leaving their workplace. At the same time, this system measures employee temperature in real time when they clock-in, which stated in Malaysia Ministry of Health (MOH) and Ministry of International Trade and Industry (MITI) Standard Operation Procedures for all organizations in Malaysia. Moreover, this new approach will eliminate the probability to get infected of COVID-19. Employee may get infected if using attendance system with shareable scanner. For example, attendance system with fingerprint scanner might spread COVID-19 left by a positive carrier. Next employee will catch the virus after touching the contaminated fingerprint scanner and then touching their eyes, nose, or mouth.

Blockchain technology will be able to secure the attendance data, thus improve the integrity of the stored data. In addition, this method will be easy to implement, as no additional devices are required. On the employee side, they need to install the developed mobile application on their smart phone. The mobile application will be used to authenticate their attendance to the workplace. While employers are required a computer storage for storing all employees' data. Therefore, the proposed contactless attendance system using face recognition, body temperature measurement and GPS on blockchain technology will overcome current issues, thus helping in preserve a healthy workplace on top of secured technology.

Acknowledgements The authors would like to express sincere gratitude to National University of Malaysia for supporting this research.

References

1. Al Sheikh R et al. (2008) Developing and implementing a barcode based student attendance system. Int Res J Eng Technol 497:1–12. Available: www.irjet.net
2. Navin K, Shanthini A, Mukesh Krishnan MB (2018) A mobile based smart attendance system framework for tracking field personals using a novel QR code based technique. Proc 2017 Int Conf Smart Technol Smart Nation SmartTechCon 2017, 1540–1543. https://doi.org/10.1109/smarttechcon.2017.8358623
3. Deugo D (2015) Using QR-codes for attendance tracking. Int Conf Front Educ CS CE, 267–273. Available: http://worldcomp-proceedings.com/proc/p2015/FEC2508.pdf
4. Elbehiery H (2019) Enhancement of QR code student's attendance management system using GPS, 21(4):18–30. https://doi.org/10.9790/0661-2104011830
5. Ayop Z, Lin CY, Anawar S, Hamid E, Azhar MS (2018) Location-aware event attendance system using QR code and GPS technology. Int J Adv Comput Sci Appl 9(9):466–473. https://doi.org/10.14569/ijacsa.2018.090959
6. Abd. Rahni AA, Zainal N, Adna MF, Othman NE, Bukhori M (2015) Development of the online student attendance monitoring system (SAMS™) based on QR-codes and mobile devices, 10:28–40
7. Kamaraju M, Kumar P (2015) Wireless fingerprint attendance management system

8. Yadav D (2015) Fingerprint based attendance system using microcontroller and labview. Int J Adv Res Electr Electron Instrum Eng 04:5111–5121. https://doi.org/10.15662/ijareeie.2015.0406029
9. Zainal NI, Sidek KA, Gunawan TS, Manser H, Kartiwi M (2014) Design and development of portable classroom attendance system based on Arduino and fingerprint biometric. In: The 5th international conference on information and communication technology for the muslim world (ICT4M), pp 1–4. https://doi.org/10.1109/ict4m.2014.7020601
10. Bhattacharya S, Nainala G, Das P, Routray A (2018) Smart attendance monitoring system (SAMS): a face recognition based attendance system for classroom environment
11. Sawhney S, Kacker K, Jain S, Singh SN, Garg R (2019) Real-time smart attendance system using face recognition techniques. In: 2019 9th international conference on cloud computing, data science and engineering (confluence), pp 522–525. https://doi.org/10.1109/confluence.2019.8776934
12. KPN Reddy, Alekhya T, Sushma Manjula T, Krishnappa R (2020) AI-Based attendance monitoring system 9
13. Fell S (2019) AI to mark attendance. Telegraph. https://www.telegraphindia.com/technology/ai-to-mark-attendance/cid/1688348. Accessed 20 Apr 2020
14. Ho CC, Eswaran C (2011) Consodilation of fingerprint databases: a Malaysian case study. Proc 2011 11th Int Conf Hybrid Intell Syst HIS 2011, pp 455–462. https://doi.org/10.1109/his.2011.6122148
15. Bitzionis T (2020) Malaysian government to implement national biometric registration system. Find Biometrics. https://findbiometrics.com/biometrics-news-malaysian-government-seeks-implement-national-biometric-registration-system-021102/. Accessed 20 Apr 2020
16. Sajadi AM (2017) Call for biometric system to screen students. The Star. https://www.thestar.com.my/news/education/2017/01/08/call-for-biometric-system-to-screen-students/. Accessed 20 Apr 2020
17. Kumar SMKVP, Kumar KK, Krishna RS, Sri PSGA (2019) Incorporation of blockchain in student management system. Int J Innov Technol Explor Eng 8(6):664–668

Identification and Detection of RTN in Short-Channel MOS Devices of Communication Power Supply

Xinxin Fan, Hongmei Yan, Xin Xu, and Hui Yao

Abstract Random Telegraph Noise(RTN) is an important parameter to characterize the short channel MOS device reliability of communication power supplies, in order to identify and detect the RTN, we adopt the noise scattering plot(NSP), combined with k order median filter to identify the noise and we propose an improved variable step size LMS algorithm for the detection. The experimental results show that the new method can effectively identify the RTN component and its related parameters, especial in smaller time constant and the same precision, the error of the parameter extraction decreased 40% than traditional methods.

Keywords RTN · NSP · LMS algorithm · Short channel MOS devices · Communication power supply

1 Introduction

At room temperature, the relative amplitude of the RTN over 60% in short-channel MOS devices of Communication power supply [1, 2]. It is the existence of noise that not only reduces the sensitivity of low frequency electronic circuit system but also coupled with phase noise, influence the high frequency digital logic device application [3, 4], which directly results in the decrease of device reliability. Thus the study of the RTN is a significant important.

The first step is to identify the RTN and then detect it, However, at present we can only identify bi-level RTN by observing the RTS time series, but for Multi-Level RTN, we can not judge through direct observation, because of the complex waveform and the interference of background noise, even cannot detect whether it has the RTN in short-channel MOS devices [5].

X. Fan (✉) · X. Xu · H. Yao
State Grid Tongling Electric Power Supply Company Tongling, Tongling, China
e-mail: Fanxxok@126.com

H. Yan
State Grid Zongyang County Electric Power Supply Company, Zongyang, China

© The Author(s), under exclusive license to Springer Nature Singapore Pte Ltd. 2021
Z. Zakaria and S. S. Emamian, *Advances in Electrical and Electronic Engineering and Computer Science*, Lecture Notes in Electrical Engineering 741,
https://doi.org/10.1007/978-981-33-6490-5_9

In this paper, we adopt the combination of NSP and median filter method to identify RTN. We propose an improved detection algorithm based on adaptive digital filter for detection.

2 Identification Method of RTN

In short-channel MOS devices, the component of Gaussian noise includes a low thermal noise, flicker noise, etc. If RTN still exists, its influence is more important than other Gaussian noise components [6]. So it is important to the recognition of the RTN. NSP method can identify RTN component in low frequency noise signal, its algorithm process is as follows (Fig. 1):

First, measure the low frequency noise signal $x(t)$, get the time series (1) through the AD conversion.

$$\{x(n)\} = \{x(1), x(2), \cdots x(2N)\} \tag{1}$$

Second, establish the relation between the matrix **X** and **Y**, classify the sequence $\{x(n)\}$ as follows:

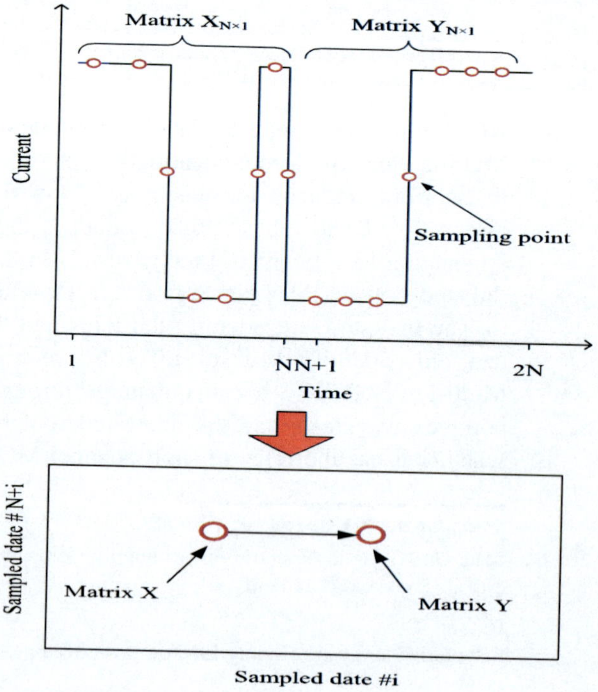

Fig. 1 The process of RTN identification

$$\mathbf{X} = \begin{bmatrix} x(1) \\ x(2) \\ \vdots \\ x(N) \end{bmatrix}, \quad \mathbf{Y} = \begin{bmatrix} x(N+1) \\ x(N+2) \\ \vdots \\ x(2N) \end{bmatrix} \tag{2}$$

Finally, draw a scatter plot of X on Y by using MATLAB or Origin.

Because the low frequency noise signal of short channel MOS devices often contains Non-Gaussian RTN, Gaussian background flicker noise and thermal noise, etc. In view of these two types of noise, to facilitate the analysis of noise components, we can divide it into three cases for discussion.

2.1 Only Non-gaussian RTN Components Exist

If the low frequency noise signal contains only the RTN. First, we assume that the low frequency noise signal contains only two level RTN, get Fig. 2 by NSP method, we can see the RTN in the scatter plot which has four similar to oval respectively around the center coordinate of (−X, −X), (−X, X), (X, −X), (X, X). In the same way, if the noise signal contains three RTN levels, the NSP diagram should be in center of eight around their approximate elliptic distribution pattern. If there are more than N value levels of the RTN, we can get the total of N2 the NSP figure which is similar to ellipse shape pattern [7].

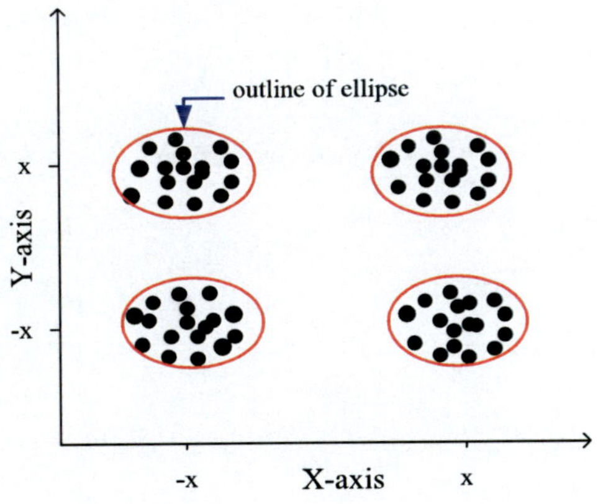

Fig. 2 RTN scatter plot

2.2 Only Gaussian Noise Composition Exist

If the low frequency noise signal contains Gaussian components of flicker noise and thermal noise, etc. the NSP graphics as shown in Fig. 3, we can only see that its scatter distribution is similar to that of ellipse.

2.3 Mixture of Gaussian Noise and Non-gaussian Noise

As shown in Fig. 4, this is a scatter diagram with Gaussian and Non-Gaussian mixed low-frequency noise. By comparing Fig. 2 with Fig. 3, it can be seen the mixed low-frequency noise that contains background noise points which resulting in mixed the intersection of scattered points in the boundary of adjacent ellipses. To eliminate the influence of these cross isolated noise points, we use K-order median filter, get sequence (3) after the data is framed.

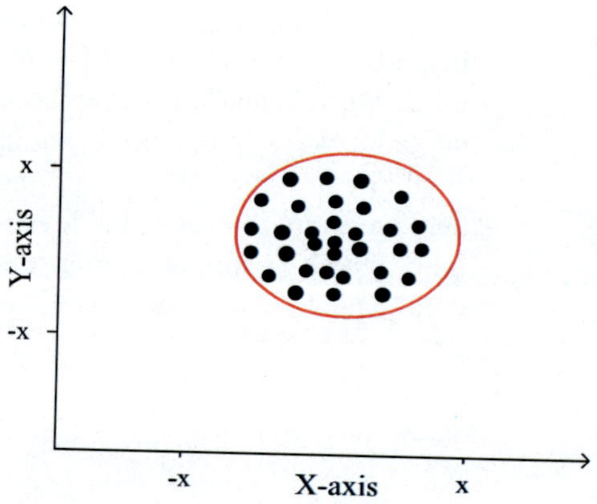

Fig. 3 1/f noise scatter plot

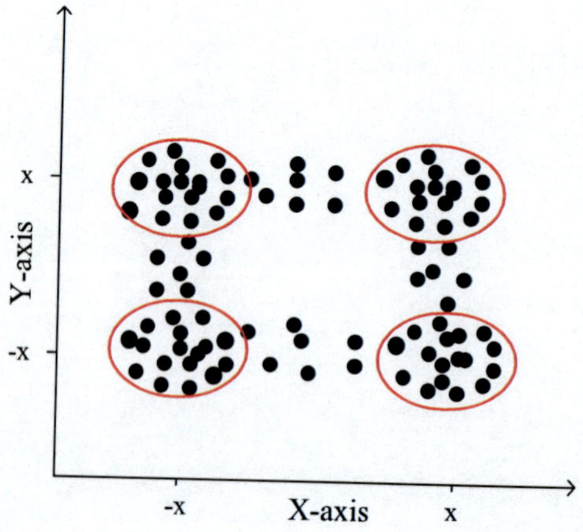

Fig. 4 Mixed noise scatter plot

$$m\{x(n)\}_k = med\{x(n-k), x(n-k), \cdots x(n+k)\}, n = k+1, 2, \ldots, 2N-k \quad (3)$$

3 Detection Method of RTN

The detection method of RTN is extraction the noise amplitude ∇ID and extraction time constant of the RTN such as trap capture ζc and emission time ζe.

3.1 Extraction of the Time Constant

The traditional time constant extraction method is to calculate the average value of rising edge, falling edge, rising edge and RTN [8, 9] ,because of the limitation of sampling points and superposition in high and low level of noise signal interference, the precision of extraction time constant is greatly restricted. To extract the adjacent two points of time series, there was related literature proposed recurrence relation [10], its concrete structure difference equations:

$$y_n = \begin{cases} x_{n+1} - x_{n-1}, & (x_{n+1} - x_{n-1})(x_n - x_{n-1}) > 0; \\ x_n - x_{n-1}, & (x_{n+1} - x_{n-1})(x_n - x_{n-1}) < 0; \end{cases} \quad (4)$$

where y_n is time series after the successive difference, x_{n+1} is sampled points at $n+1$ moment, x_n is Sampled points at n moment, x_{n-1} is Sampled points at n − 1 moment.

Because the RTN signal from the actual test often contains three parts, including low frequency interference from thermal noise, high frequency disturbance of flicker noise and RTN components. But the accuracy of extraction has been curbed. Therefore, in this paper, the LMS algorithm is improved, and a variable step LMS algorithm is proposed. Its design idea is as follows:

According to the expression of the adaptive filter [11]:

$$\mathbf{W}(n+1) = \mathbf{w}(n) + \mu\mathbf{u}(n)\, e(n) \quad (5)$$

$$e(n) = d(n) - \mathrm{w}^H(n)\,\mathbf{u}(n) \quad (6)$$

where $\mathbf{W}(n)$ is the weighted coefficient of filter, $\mathbf{u}(n)$ is the differential input signal, $e(n)$ is the relative value error, μ is to control the disturbance of the step length.

The traditional LMS algorithm cannot optimize convergence speed and stability at the same time, due to the error signal RTN have the strong correlation and weak cross correlation caused by the traditional LMS algorithm which is sensitive to noise. To solve the problem of traditional LMS algorithm, the correlation between step function and error expression is established to reduce the sensitivity of LMS algorithm to noise. The specific process is as follows:

According to the discrete expressions of correlation function (7), It is concluded that the correlation function (8), we know the index has the nature of explosive growth, to make $R_e(k)$ minimum, power exponential function $\exp|e(n)e(\mathrm{n}-1)|^{\mathbf{k}}$ can be introduced as denominator of expression, at the same time, we should consider the convergence speed of step length, so a variable α is introduced, the power exponential function is complemented to speed up the convergence speed. Final, build function expressions such as (9)

$$R_e(1) = E\{e(n)e(n-1)\} \tag{7}$$

$$R_e(k) = E\{e(n)e(n-k)\} \tag{8}$$

$$\mu(n) = \left(a - \frac{1}{\exp|e(n)e(n-1)|^b}\right) \tag{9}$$

where $R_g(1)$ is the cross-correlation function of sampling interval at 1, $R_g(k)$ is the cross-correlation function of sampling interval at k, $E\{.\}$is expectation, $e(n-k)$ is the relative value error at $n-k$, a is a constant definition and b is need to measure the optimal constant.

The specific ideas of the algorithm are:

In the beginning of convergence, to accelerate the convergence speed, we take the big step length. As the error is reduced gradually, take the small step length, to achieve small steady state low-offset noise. The introduction of factor $e(n)(n-1)$ can eliminate the error signal of weak cross correlation and strong correlation between each other, thus inhibiting the phenomenon which traditional LMS algorithm is sensitive to noise, from Fig. 5 shows that when the equation of b = 1, in the initial stage of Fig. 5, step down quickly, when the error close to zero, step length is still obvious change, do not have slowly changing characteristics. When the equation of b = 2, it improves the convergence rate of $\mu(n)$ at the condition of b = 3, and overcomes the disadvantage of a big step change at the condition of b = 1. Therefore makes the algorithm has small steady state low-offset noise, thus to achieve the best filtering performance.

3.2 Extraction of the Amplitude Parameter

Traditional method of amplitude parameter extraction [12–14] is to measures the relative magnitude of the RTN $\nabla I_D/I_D$. But under the same bias, we need to measure the absolute amplitude of the RTN ∇I_D and channel current I_D, there is a problem that under the same bias circuit condition it is hard to measure direct current and alternating current at the same time, this will brings great inconvenience and error at the extraction of the amplitude.

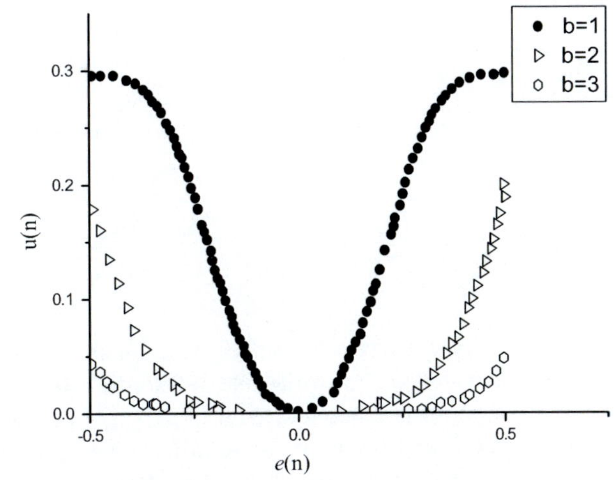

Fig. 5 Dependence of k and $\mu(n)$ also $e(n)$

Therefore, the positive and negative to measure absolute magnitude is adopted, using the transfer characteristic curve Under the condition of the same gate bias, Finally, we establish ∇I_D about I_D the graph, and then extract the amplitude parameter.

4 Experiment and Result Analysis

The samples of the experiment are SMIC 90 nm feature sizes CMOS devices, using the technology of LDD and Silicide technology, etc. The threshold voltage VT = 0.2 V, junction depth is 0.02 μm, the injection depth of LDD is 30 nm, the effective thickness of gate dielectric is 1.6 nm, the channel size NMOS single pipe is 0.18 × 0.16 μm.

4.1 Identification Verification and the Result

As shown in Fig. 6, this is the experiment used in the system of measuring short-channel MOS device, the type of Semiconductor Device Analyzer is Agilent 1,500A, WGFMU is waveform generator and fast measurement unit and RSU is remote sensing and switch unit to deliver our noise data, the DUT is our samples of short channel MOS. NSP method is used to draw the sample point for 10,000 points, the results is shown in Fig. 7 (b), after a median filter, we can get Fig. 7 (c). It can be seen the filter can eliminate noise isolated point and the influence of background noise from the RTN.

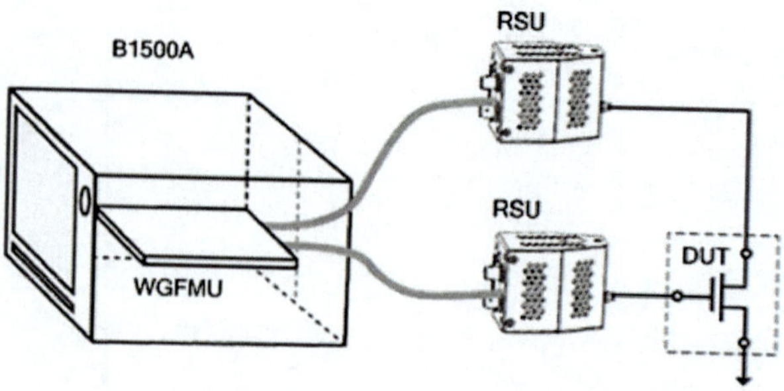

Fig. 6 The RTN measurement system

According to the analysis of the above conclusion, combined with Fig. 7 (c), we can not only identify the RTN from the sample, but also determine the level of RTN, it is consistent with the Fig. 7 (a), we can draw the basic information of the RTN, and the points around the nine center is in non-uniform distribution.

There are nine central coordinates of RTN have uneven density, we can preliminary judgment ζc is not equal with the ζe, the specific reason may about the different trap position or level in the sample inside.

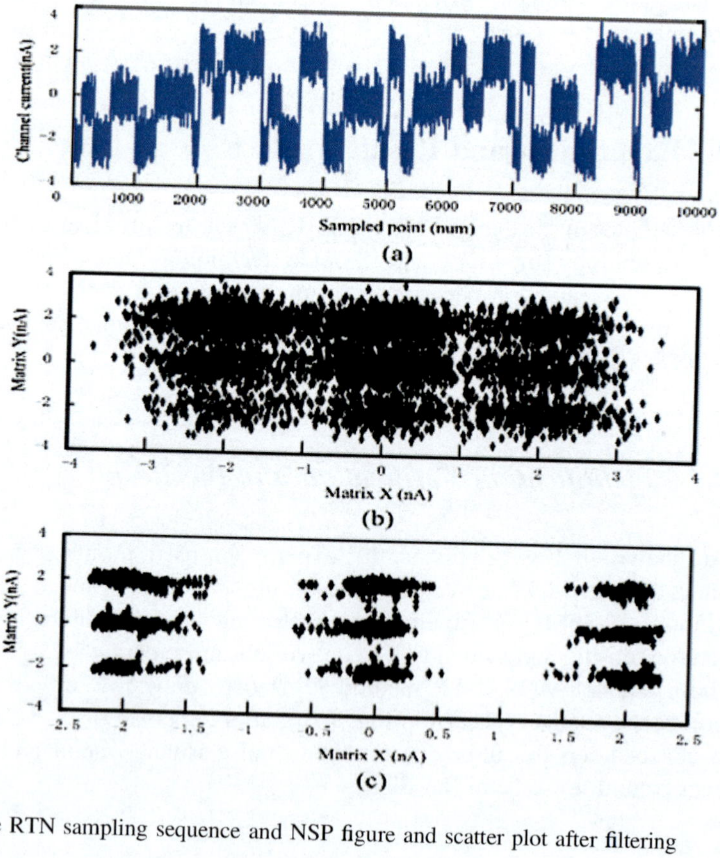

Fig. 7 The RTN sampling sequence and NSP figure and scatter plot after filtering

4.2 The Extraction of RTN Time Parameter

To extract the time parameters, first we design variable step LMS digital filter, define a = 1, and take b = 2, set up the drain voltage is 320 mv, grid voltage range from 400 to 800 mv, Sampling rate selected for 100 Kps, Sampling accuracy for 14 bit, Continuous sampling for each grid bias at point 30 s. We can get trap capture time and emission time respectively, the relationship between the gate voltages as shown Fig. 8, this is consistent with the results of literature research and satisfied with the exponential distribution. At the same time, monte carlo method and difference equation method are used to simulate the traditional extraction and literature method respectively, the detailed simulation result as shown Fig. 9, from the figure we can see the new method of error performance is superior to the traditional method and the difference equations method, compared with the traditional average algorithm, under the same precision, sampling point averages about 1/10 of the traditional method. When the time constant of trap capture and emission is relatively small, the advantage of the new method is more prominent, the error can be reduced by 40% within the number of sampling 10 points, can significantly improve the precision of parameter extraction. Compared with the modified LMS algorithm and the relevant literature proposed using finite difference equation method [2], the result follows the same trend. While in the small time constant, the modified LMS algorithm is better than difference equation method, it can increase 10% error precision. When the time constant is higher, it can increase 5% error precision.

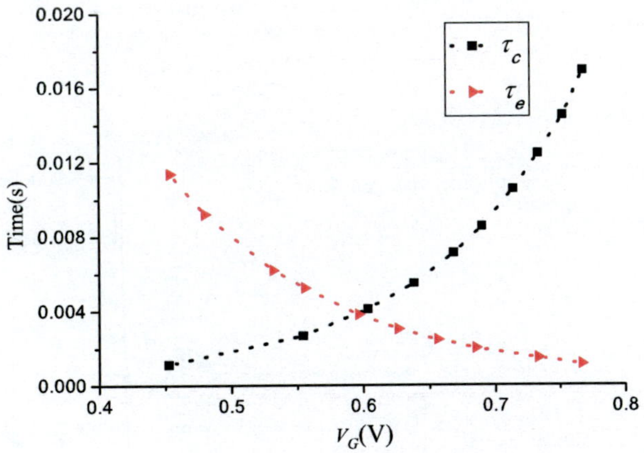

Fig. 8 Dependence of RTS time constant and gate bias

Fig. 9 The monte carlo simulation

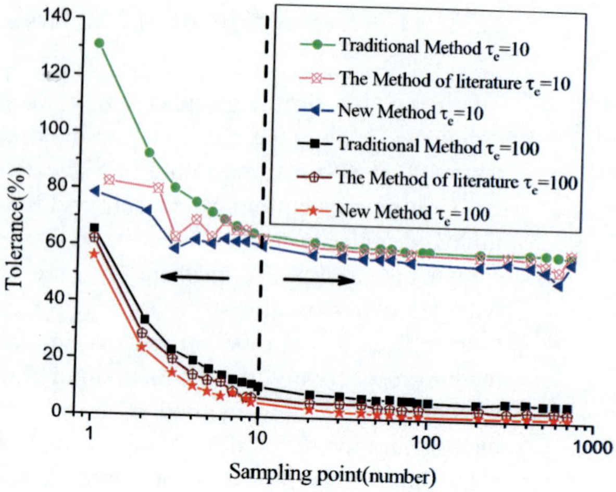

4.3 The Extraction of RTS Amplitude Parameter

To extract the amplitude parameter, changing the gate voltage range from 0 to 280 mV, continuous sampling for each grid bias at point 200 s, Fig. 11 is RTN amplitude curves under positive and negative bias conditions, combined with the transfer characteristic curves as shown Fig. 11, according to the value of any point of V_G, between Figs. 10 and 11, we can determine the relationship between I_D and ∇I_D. Meanwhile compare with mathematical model from the literature [12], we can clearly a draw conclusion that the new way to measure the margin of RTN is identical with the calculation results of the literature. Because of the expression (10) which is derived in the premise of exponential distribution, so we can infer that the new method to get the margin of RTN also accords with exponential distribution.

Fig. 10 Dependence of RTS amplitude and gate bias

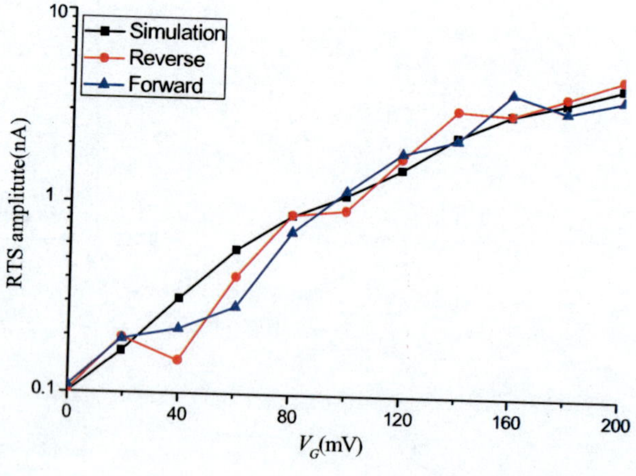

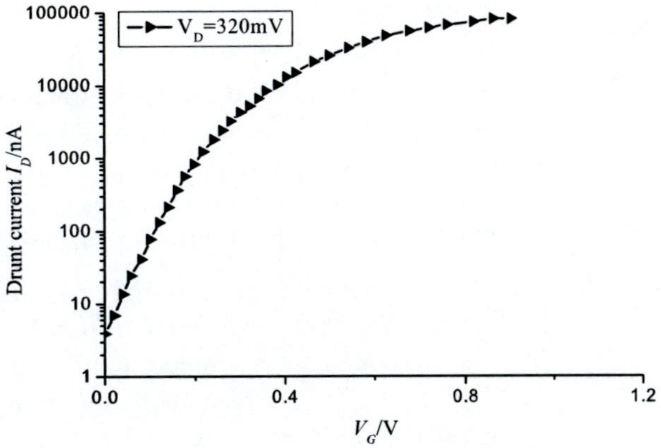

Fig. 11 Transition characteristic of the sample with the bias of V_D = 310 mv

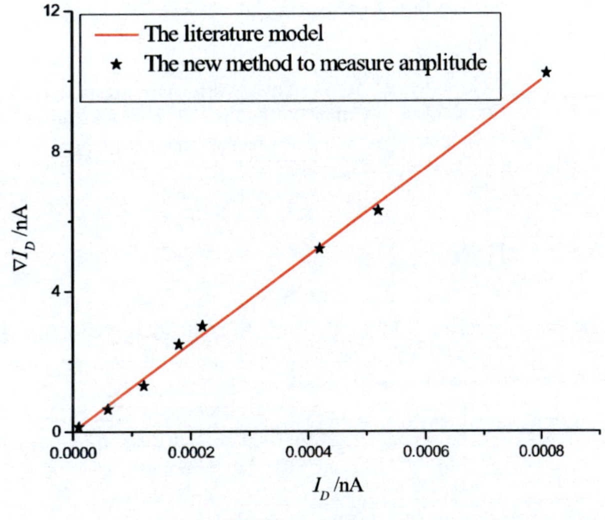

Fig. 12 Literature model and the new way to measure results

$$\Delta I_{\mathrm{D}} = \frac{4r_c^2}{WL - 2r_cL + 4r_c^2} I_{\mathrm{D}} \tag{10}$$

where W is the channel width, L is the channel length, r_c—is the distance between trap charge and channel (Fig. 12).

5 Conclusions

In this paper, NSP method and k-order median filter are used to identify RTN, It can judge the basic information such as RTN level value and center position of noise point. At the same time according to the error of relative value $e(n)$ $e(n - 1)$, a improved LMS algorithm is proposed for the detection. The experimental measurements show that the following conclusions:

(a) The new method is in conformity with the mathematical model in the process of amplitude parameter extraction from RTN.
(b) The new method can reduce 40% under the same precision error,Compared with the traditional extraction algorithm of RTN time parameter.
(c) Improved error accuracy by 10%, Compared with the related literature proposed by difference equation.
(d) The new method solved the shortcomings of traditional algorithm existed in sensitive to noise.

Acknowledgements This work was supported by a grant from National Natural Science Foundation of China (NSFC 61271115). The authors would like to thank Xinxin Fan for excellent technical support and Professor Xiaojuan Chen for critically reviewing the manuscript.

References

1. Kola SR, Li Y, Thoti N (2020) Random telegraph noise in gate-all-around silicon nanowire MOSFETs induced by a single charge trap or random interface traps. J Comput Electron 19 (1):12–16
2. Fan XX, Yang LY, Chen XG et al (2018) Detection and analysis of random telegraph signal noise in P-MOSFET. Dian zi Jishu Yingyong 44(8):44–46
3. Möhle C, Zota CB, Hellenbrand M et al (2017) 1/ f and RTN in InGaAs Nanowire MOSFETs. Microelectron Eng 36(6):52–55
4. Cai XJ (2020) Quantum depraving induced by non-markovian random telegraph noise. Sci Rep 10(1):18–20
5. Both TH, Croon JA, Silva MD et al (2017) Autocorrelation analysis as a technique to study physical mechanisms of MOSFET low-frequency noise. IEEE Trans Electron Devices 18 (6):1–8
6. Fried JP, Bian XY, Swett JL et al (2020) Large amplitude charge noise and random telegraph fluctuations in room-temperature graphene single-electron transistors. Nanoscale 12(2):26–29
7. Ishida HK, Kagawa KC, Takashi ZB et al (2018) Multi-aperture-based probabilistic noise reduction of random telegraph signal noise and photon shot noise in semi-photon-counting complementary-metal-oxide-semiconductor image sensor. Sensors 8(4):26–32
8. Li Z, Sotto M, Liu FY et al (2018) Random telegraph noise from resonant tunneling at low temperatures. Sci Rep 8(1):4–9
9. Yian YC, Potirakis SM, Stavrinides SG et al (2019) Intermittency-induced criticality in the random telegraph noise of nanoscale UTBB FD-SOI MOSFETs. Microelectron Eng 16 (8):216:219
10. Martin J, Rodrigue R, Nafria M et al (2014) New weighted time lag method for the analysis of random telegraph signals. IEEE Electron Device Lett 35(4):479–481

11. Ban M (2020) Decoherence of a two-quite system interacting with initially correlated random telegraph noises. Quant Inf Process 19(2)
12. Saraza PC, Martin JM, Castro RL et al (2019) A detailed study of the gate/drain voltage dependence of RTN in bulk NMOS transistors. Microelectron Eng 19(8):215–217
13. Awano H, Hiromoto M, Sato T (2015) An efficient method for calculating RTN-induced failure probability of an SRAM cell. Design Autom Test 16(2):346–349
14. Markus H, Pekka O, Johannes K et al (2019) Low-frequency noise in nanowire and planar III-V MOSFETs. Microelectron Eng 26(16):215–220

Evaluation of Interactive WebSIR Using Software Usability Measurement Inventory (SUMI)

Mohd Suffian Sulaiman and Azri Azmi

Abstract The classical studies of information retrieval such as precision, recall, accuracy and F-measure, detached the human out from the model of evaluation. As this study have proposed a model that would engage with the user, the assessment must also take into account the experiences that exist between users and systems. In this paper, the Software Usability Measurement Inventory (SUMI) is used to evaluate a semantic image retrieval application namely as WebSIR. WebSIR is a web-based application that developed using Java programming language, Liferay and Jena. A collection of questions derived from the SUMI instrument were used. The results demonstrated that 85% evaluation test of respondents are satisfied and they considered this application is easy to be used.

Keywords SUMI · Liferay · Semantic image retrieval · Interactive information retrieval

1 Introduction

The evaluation is performed to further evaluate the efficiency of the program proposed. A good information retrieval system has a better strategy and offers a better classified list to satisfy the information needs of the user. According to the Cranfield paradigm, a document collection (the document may be in the form of articles or images), a set of queries (or topics) and relevant judgements identified for the document collection are needed to determine the quality of the retrieved results provided by a retrieval system [1]. Generally, the evaluation performance for

M. S. Sulaiman (✉)
Faculty of Computer and Mathematical Sciences, Universiti Teknologi MARA (UiTM), Shah Alam, Selangor, Malaysia
e-mail: Suffian@tmsk.uitm.edu.my

A. Azmi
Razak Faculty of Technology and Informatics, Universiti Teknologi Malaysia (UTM), Kuala Lumpur, Malaysia

© The Author(s), under exclusive license to Springer Nature Singapore Pte Ltd. 2021
Z. Zakaria and S. S. Emamian, *Advances in Electrical and Electronic Engineering and Computer Science*, Lecture Notes in Electrical Engineering 741,
https://doi.org/10.1007/978-981-33-6490-5_10

information retrieval studies such as precision, recall, F-measure and accuracy were used [2, 3]. This form of evaluation, however, is still insufficient to determine whether the consumers are pleased with using it before the program can be handed over to the consumer. In other words, the classical evaluation of information retrieval usually asks the question, can this program get relevant documents? While the interactive evaluation of information retrieval raises the question of whether people should use this program to access relevant documents [4].

2 Related Work

The user acceptance is very critical otherwise the program or product produced does not make any sense for the users. Therefore, an evaluation must be carried out before a product can be delivered to the consumer. 24 questionnaires evaluations can be used since the late 1980s until today and five of them have been shortlisted based on the highest number of quality criteria as shown in Table 1 [5]. The five questionnaires are Questionnaire for User Interface Satisfaction (QUIS) [6], Usefulness, Satisfaction and Ease of Use Questionnaire (USE) [7], Software Usability Measurement Inventory (SUMI) [8], Technology Acceptance Model questionnaire (TAM) [9] and Purdue Usability Testing Questionnaire (PTUQ) [10].

The QUIS is a method designed to measure subjective satisfaction of users with particular aspects of the interaction between the person and computers.

Table 1 Usability criteria

Usability criteria		QUIS	TAM	SUMI	PUTQ	USE
ISO 9421-11 criteria [11]	Effectiveness	X		X		X
	Efficiency		X	X		
	Satisfaction	X		X		X
ISO 9241-112 criteria [12]	Detectability	X	X	X		
	Discriminability	X		X	X	
	Appropriateness			X	X	
	Consistency	X		X	X	X
	Comprehensibility	X	X	X	X	
Scapin and Bastien criteria [13]	Guidance	X		X	X	X
	Workload			X	X	X
	Explicit control			X	X	
	Adaptability	X	X	X	X	X
	Error management	X		X	X	X
	Consistency	X		X	X	X
	Significance of codes	X		X	X	
	Compatibility			X	X	

A multi-disciplinary research team at the University of Maryland Human-Computer Interaction Lab, created it in 1987. The QUIS is currently available in version 7.0 with a demographic questionnaire, a 6-scale indicator of overall device efficiency and 9 different design factors. The USE measures a product or service's subjective usability. It has a 30-item survey that examines four usability dimensions: utility, ease of use, ease of learning and satisfaction. This metric applies to different usability assessment scenarios because it is non-proprietary and technology-agnostic. Items in the USE also have good face validity with a clear and appropriate description. The SUMI is a rigorously tested and validated tool for assessing the usability of a computer from end-users. SUMI also used to assess the user experience when it first appeared in the 1990s. The TAM was one of the most prominent technologies acceptance models, with two key factors shaping an individual's intent to use modern technology: perceived user-friendliness and perceived usefulness while the PTUQ was developed at the University of Purdue, based on eight human factors related to the usability of the program. It is derived from the experimental and theoretical basis outlined in the published literature. PTUQ has strong construct validity and content validity. The usability questionnaire will eventually be used to extract usability index measurements for each application. Thus, the emphasis on the user-friendly design of computer software systems would be further increased.

Based on five shortlisted questionnaires, the SUMI was selected to evaluate the program because it was rigorously tested and proven method of measuring software quality from the end user's point of view and fulfil all the usability criteria as shown in Table 1 that adapted from Assila et al. (2016). A set of questions that are taken from SUMI instrument was used. Compared to other evaluation questionnaires, SUMI cover all the usability criteria such as effectiveness, efficiency, satisfaction, detectability, discriminability, appropriateness, consistency, comprehensibility, guidance, workload, explicit control, adaptability, error management, consistency, significance of codes and compatibility.

3 Methodology

WebSIR is an application developed by using Java programming language, Liferay enterprise portal technology [14] and Jena toolkit [15] that provide an application programming interface (API) for Resource Description Framework (RDF) [16]. It has four functional requirements which are input text query, view the initial result, select related properties and view the filtered result as illustrated in Fig. 1 [17]. The use case diagram showing a series of use cases and actors and their relations; use case diagrams describe a system's static use case view [18]. In WebSIR, the user needs to enter a natural language query inside the search text box. Then, the user can view the initial result once the submit button clicked. Next, the user needs to mark related properties to filter the initial result that is dynamically displayed with the initial queried results. The dynamic means the list of properties can flexibly

Fig. 1 Use case diagram of WebSIR

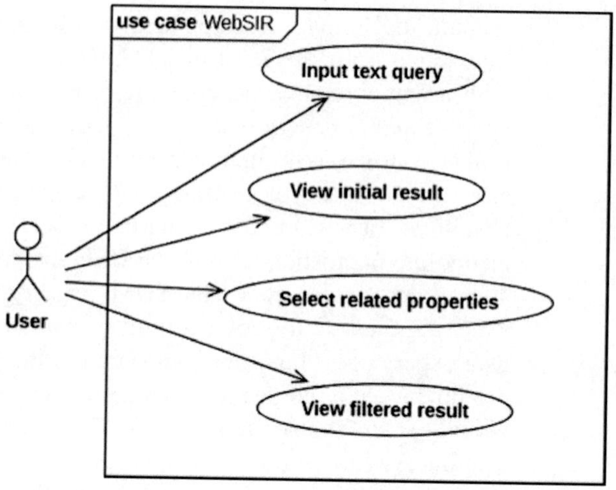

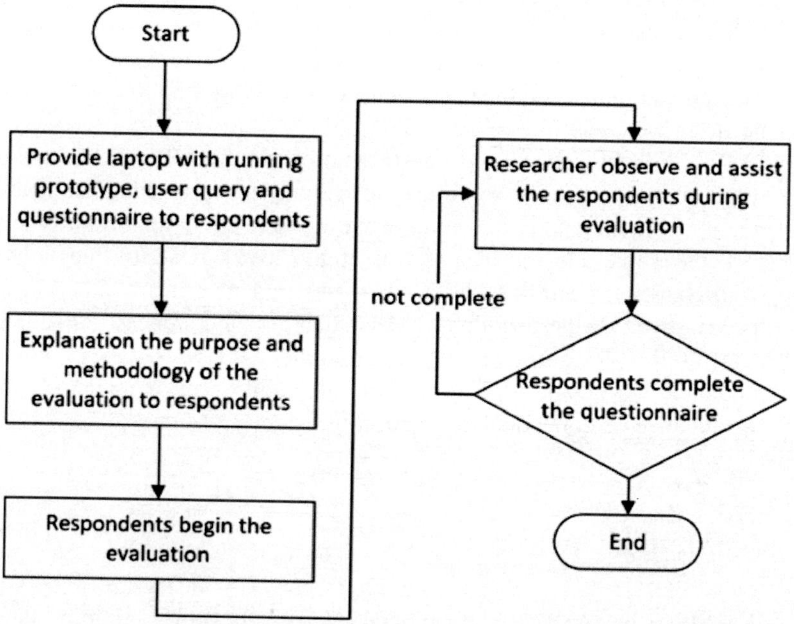

Fig. 2 Flow chart of SUMI evaluation

change depending on the retrieval query. Then, the marked properties are fed back into the system as a new redefined query for the following cycle of the retrieval process to obtain the better queried result [19].

The usability evaluation of the program was done by meeting the respondents physically and the flow of SUMI evaluation can be illustrated in Fig. 2. During the meeting, the laptop with running program, user query and the questionnaires are provided. The researcher met with the respondents for 10 min to explain the purpose of the measurement and explain the methodology of SUMI evaluation. The respondents received the verbal instructions from the researcher, throughout the evaluation session. The researcher was present to help with any difficulties with the questionnaire and to answer questions asked from respondents. For next phase the users were asked to complete the SUMI questionnaire for user-interaction satisfaction. The evaluation sessions lasted around 20 min each.

The program and questionnaire were tested by ten individuals. All of them are an undergraduate student from Faculty of Applied Sciences, Universiti Teknologi MARA (UiTM), Shah Alam, Selangor, Malaysia. As part of the evaluation process, we ensured that all respondents had some basic knowledge to search the herbal medicinal plant information using a web browser.

4 Result and Discussion

The 20 SUMI questionnaires were selected from 50 items which the respondents need to select one of three responses ("Agree", "Undecided" and "Disagree") [20]. For the analysis purpose, a certain value is assigned for each scale as shown in Table 2. The average is collected from the summation of input from the respondents. If the result is between 1.00 and 1.99, it shows the user agreed with the specific measurement of the program. If the result is between 2.00 and 2.99, it shows the user is undecided and if the result is between 3.00 and 4.00, it shows the user disagrees with the program. Table 3 shows the details of the result.

Table 2 Attribute in the evaluation form

Scale	Agree	Undecided	Disagree
Value	1.00–1.99	2.00–2.99	3.00–4.00

Table 3 Result of evaluation test

No.	Questions	Result (Average)
1	The software reacts quicker to input	1.17
2	The directions and prompt are beneficial	1.12
3	Simple to learn how to use the software	1.16
4	The support information is really helpful	1.2
5	Spend less time mastering the software	1.17
6	The software is compliant	1.18
7	Information presented is transparent and comprehensible	2.45
8	Enough on-screen details if needed	2.57
9	The software is reliable	1.14
10	Can understand and act on the supplied information	1.12
11	The software is not awkward	2.44

(continued)

Table 3 (continued)

No.	Questions	Result (Average)
12	Less to learn before you can use the software	1.2
13	Operation can be carried out right away	1.28
14	The software is not frustrating	1.24
15	The software had fixed the problem	1.21
16	Menu structure is rational	1.13
17	Can be economical of keystroke	1.23
18	Few required action	1.16
19	Simple to make software exactly what you want	1.15
20	The software provided correct results	1.24

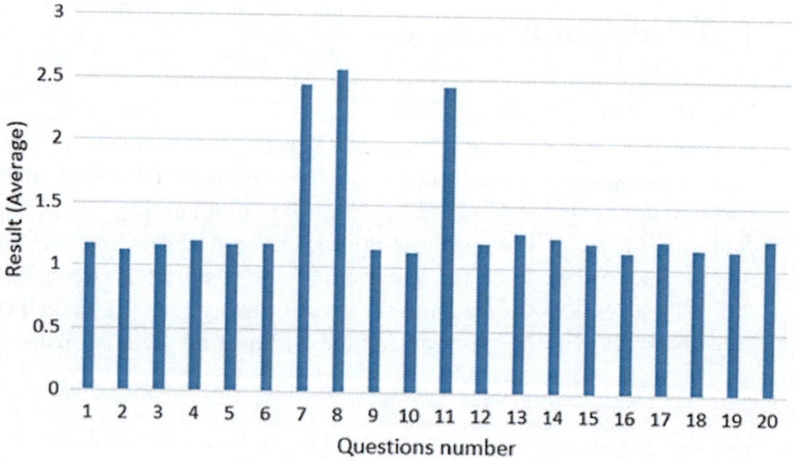

Fig. 3 Result of evaluation test graph

Figure 3 shows the respondent's feedback that can be categorized into two categories namely as agree and undecided. No feedback goes to disagree category. From the 20 questions, the respondents could not indicate the feedback in three questions which are information presented is transparent and comprehensible, enough on-screen details if needed and the software is not awkward. This is because the program dealing with the variety of datatype properties, which the respondents need to be understood before they can proceed with the searching enhancement process.

5 Conclusions

This paper presents interactive WebSIR evaluations using SUMI. Overall, the results show that the respondents are satisfied with the program and they considered it possible to use the program even though the issue of bias can be happening when involving with human interaction. The result indicates that 85% evaluation test are satisfied by respondents. For future study enhancement, the coverage of metadata can be extended not only for the public user but to the entire related user. It will provide the real user in this domain with a more comprehensive dataset.

Acknowledgements The authors gratefully acknowledge the funding support received from Faculty of Computer and Mathematical Sciences, Universiti Teknologi MARA (UiTM), Malaysia.

References

1. Jones KS, Willett P (1997) Readings in information retrieval, 1st edn. Morgan Kaufmann Publishers, Inc.
2. Del Bimbo A (1999) Visual information retrieval, 1st edn. Morgan Kaufmann, California, USA
3. Manning CD, Raghavan P, Schutze H (2008) An introduction to information retrieval. Cambridge University Press, New York, USA
4. Kelly D (2009) Methods for evaluating interactive information retrieval systems with users 3 (1–2)
5. Assila A, De Oliveira K, Ezzedine H (2016) Standardized usability questionnaires: features and quality focus. J Comput Sci Inf Technol 6(1):15–31
6. Chin JP, Diehl VA, Norman KL (1988) Development of an instrument measuring user satisfaction of the human-computer interface. Conf Hum Factors Comput Syst Proc, Part F1302:213–218. https://doi.org/10.1145/57167.57203
7. Lund AM (2001) Measuring usability with the USE questionnaire. Usability Interface 8(2):3–6. https://doi.org/10.1177/1078087402250360
8. Kirakowski J, Corbett M (1993) SUMI: the software usability measurement inventory. Br J Educ Technol 24(3):210–212. https://doi.org/10.1111/j.1467-8535.1993.tb00076.x
9. Davis FD (1989) Perceived usefulness, perceived ease of use, and user acceptance of information technology. MIS Q 13(3):319–340
10. Lin HX, Choong YY, Salvendy G (1997) A proposed index of usability: a method for comparing the relative usability of different software systems. Behav Inf Technol 16(4–5):267–277. https://doi.org/10.1080/014492997119833
11. ISO 9241-11 (1998) Ergonomic requirements for office work with visual display terminals (VDTs): guidance on usability
12. ISO 9241-112 (2017) Ergonomics of human-system interaction—Part 112: principles for the presentation of information
13. Bastien JMC, Scapin DL (2006) Ergonomic criteria for the evaluation of human-computer interfaces to cite this version: HAL Id: inria-00070012
14. Yuan JX (2010) Liferay portal 6 enterprise intranets. Packt Publishing
15. McBride B (2002) Jena: a semantic web toolkit. IEEE Internet Comput 6(6):55–58
16. Sulaiman MS, Nordin S, Jamil N (2017) An object properties filter for multi-modality ontology semantic image retrieval. J Inf Comun Technol 16(1):1–19
17. Sulaiman MS, Nordin S, Jamil N, Halin AA (2014) Multi-modality ontology for herbal medicinal plant semantic based image retrieval. In: Knowledge Management International Conference (KMiCE)

18. Sulaiman MS, Ghani MKA (2011) Software requirement, analysis and design. Object Oriented Approach UML, Penerbit UTeM
19. Sulaiman MS, Nordin S, Jamil N (2015) Enhancing the performance of multi-modality ontology semantic image retrieval using object properties filter. In: International conference on computing and informatics (ICOCI)
20. Mansor Z, Kasirun Z, Yahya S, Arshad N (2012) The evaluation of webcost using software usability measurement inventory (SUMI). Int J 2(2):197–201

Raspberry Pi Based Wireless Interface System for Automated Microfabrication in the Context of Industry 4.0

Yih Bing Chu and Weng Kent Yap

Abstract The context of Industry 4.0 emphasizes on networking of manufacturing stations to a remote server for process planning and visualization. Especially for electronics production which is sensitive to process variations, networking of the manufacturing stations facilitates the monitoring of manufacturing stations remotely. In contrast with manual look up of the machines, the aforementioned approach allows prompt notification in real-time to manufacturer for immediate resolution and as a result, breakdown time of the production can be minimized. Therefore, this paper presents a wireless interface system for electronics microfabrication stations. Each microfabrication stations are interfaced with a WiFi module for wireless data transmission to a remote server established on Raspberry Pi. The experimental setup and testing of the developed system is presented in this paper for public interest.

Keywords Industry 4.0 · Microfabrication · Photolithography · Raspberry Pi · Wireless interface system

1 Introduction

In current electronics manufacturing industry, patterns of the devices are transferred using photolithography techniques. Association of various stations such as coating stations to deposit photoresist film on wafer, ultraviolet lighting station for transferring the device pattern onto the photoresist film and etching station to remove the excessive metal layers are essential to form the final electronic devices [1]. However, often due to the small aspect ratio of the metal layers, the dimension or feature size of the fabricated device is often subjected to unwanted deviation during the entire microfabrication processes. Hence, networking of the stations are

Y. B. Chu (✉) · W. K. Yap
Tunku Abdul Rahman University College, Kuala Lumpur, Malaysia
e-mail: allan.chu.y.b@gmail.com

© The Author(s), under exclusive license to Springer Nature Singapore Pte Ltd. 2021

Z. Zakaria and S. S. Emamian, *Advances in Electrical and Electronic Engineering and Computer Science*, Lecture Notes in Electrical Engineering 741,
https://doi.org/10.1007/978-981-33-6490-5_11

required to communicate the process discrepancies between each other and coordinate the production of electronic devices systematically.

In accordance to the current industry revolution, the interconnection of machines to a centralized network facilitate the transfer and processing of production data. In this context, the remote server is responsible to interpret, control and provides data visualization of the manufacturing process to interested parties outside of the manufacturing stations. This mechanism assists manufacturer to manage more effectively the fabrication compared to manual look up of the production machine and in addition allows real-time notification for prompt action should breakdown occurs. Despite the inception of Industry 4.0, the development of such system especially for the aforementioned microfabrication process has yet been reported. In view of this, the development of such system is both strategic in time and advantageous. Therefore, the objective of this research is to design an interconnected interface between the microfabrication stations which allows the data to be communciated to a remote server for real-time visualization. In this work, WeMos WiFi ESP8266 is used to implement the wireless interface between the production stations and Raspberry Pi microcontroller is applied as the embedded server for remote visualization. The microfabrication stations should transfer the production status and able to be visualized at the remote server.

The remainder of this paper is structured as follows: Sect. 2 outlines the proposed methodology; Sect. 3 describes the initial implementation of the system and an evaluation of the proposed system; Sect. 4 draws the conclusion, limitation and future improvement of the work.

2 Methods

In electronics manufacturing, most devices are patterned in two steps; patterning of photoresist film on top of the functional material and transferring the photoresist pattern onto the functional material. Photolithography has been the forerunner among the patterning techniques for microfabrication of miniature devices such as microelectronics, micro devices and micro-electro-mechanical-system (MEMS). The technology involves making a mask and then project the mask pattern via ultraviolet (UV) light onto a photoresist-coated wafer. Lastly, the wafer is developed and etched to remove the excessive photoresist and metal layers respectively and form the final electronics pattern. The first portable and low-cost photolithography station was reported by Huntington et al. in 2011 however for small scale production and laboratory use [2]. The configuration recorded a resolution of 200 nm feature size at the consumption of 0.2% power of the mercury-vapor lamps used in the industrial mask aligner system. Similarly, a portable UV lighting station will be constructed along with tailor-made microfabrication stations. For simplicity, commercial 4 inches copper board readily coated with copper layers will be used for testing of the proposed system. Figure 1 shows the general microfabrication processes for microfabrication of electronic devices.

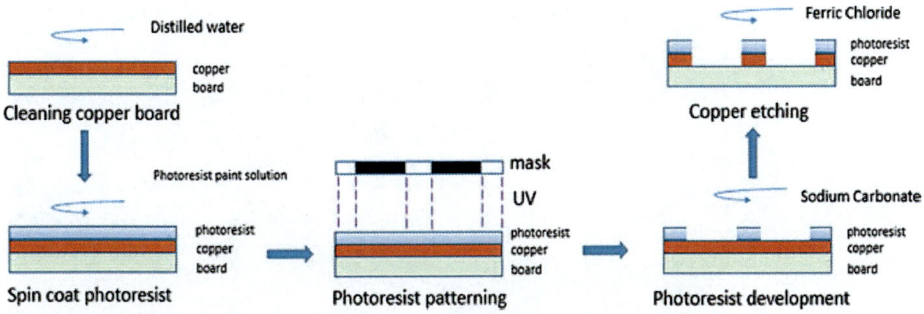

Fig. 1 Microfabrication processes for electronic devices

In this work, WiFi module ESP 8266 is applied as communication interface of the stations with the remote server. The WiFi module utilizes Message Queuing Telemetry Transport (MQTT) data transfer protocol which is suitable for light-weight data transmission and wireless sensing application [3]. In recent years, the WiFi module has been used in application such as smart home monitoring [4], heart rate monitoring [5] and monitoring of oyster mushroom cultivation [6]. The proposed system consists of spin coater for cleaning and coating of photoresist onto the copper board, UV lighting station to form patterns onto the coated photoresist layer and robotic arm handler to transfer the copper board between the spin coater and UV lighting station. All of these stations are installed with Wemos and communicated to a remote computer (Raspberry Pi). In this work, the spin coater station is applied to perform cleaning of the copper board, coating of photoresist and wet etching. The experiments and testing of the proposed system are conducted in room

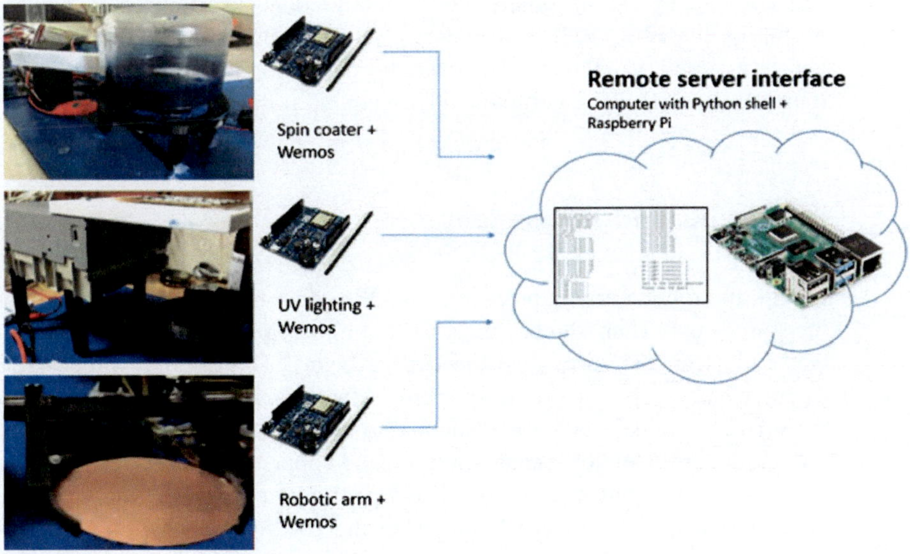

Fig. 2 Design of the proposed system

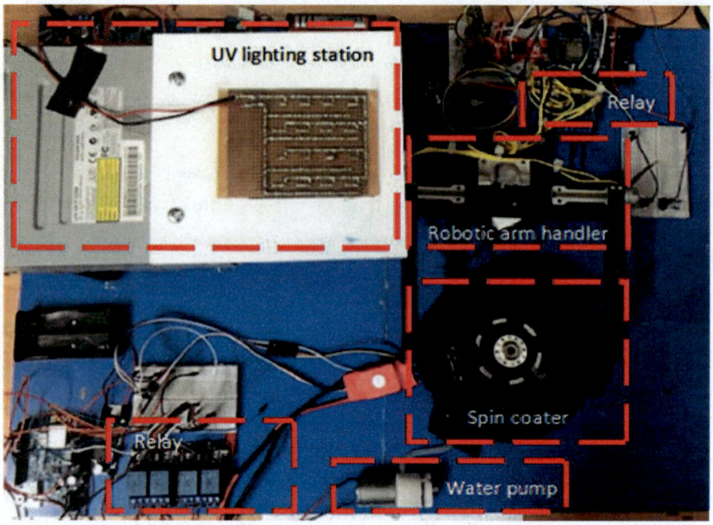

Fig. 3 The proposed setup (top view); spin coater removed casing

temperature condition. Figures 2 and 3 shows the design and overall setup of the proposed system respectively.

The spin coater is built-in with encoder sensor (tcrt5000 optical sensor) to determine the speed of rotation. The encoder will then feedback the speed measurement to the motor to spin till a desirable speed is achieved. Such system is also termed as close-loop control system. In this work, 1,000 rpm is set as default spinning speed for all the processes under the spin coater operation. For the UV patterning station, 16 pieces of UV LED (each connected in series with a 100 Ohm resistor) is aligned in square fashion and connected to a 5 V power supply. Similarly, a VEML 6070 UV intensity sensor is built-into the station to feedback the desireable UV intensity. The mask for patterning is printed in black colour on a transparent sheet using commercially available printer.

3 Results and Discussions

Firstly, the robotic arm handler transfers the coper board from its initial position to the spin coater. Then, the nozzle at the top of the spin coater casing sprays distilled water to clean the surface of the copper board. The cleaning process ends after drying of the copper board via spinning. After this, photoresist paint is dipped onto the surface of copper board and spinned to form the photoresist dry film. After that, the robotic arm handler transfers the cleaned copper board to UV lighting station to form the desired pattern onto the coated photoresist film. Subsequently, the robotic handler transfers the copper board from the UV lighting station back to the spin

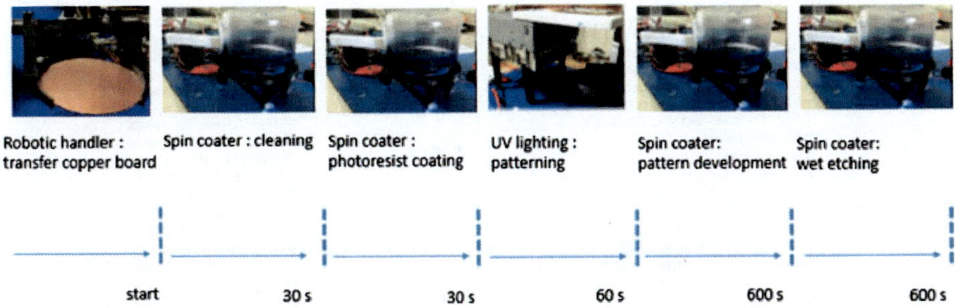

Fig. 4 Timing sequence of the overall processes

Fig. 5 Patterning mask (left); Copper board after development (mid); copper board after etching (right)

coater. The spraying cum spinning process continues at the spin coater with developer solution (1 g sodium carbonate per 1,000 ml distilled water) and copper etchant (ferric chloride) to form the final electronic patterns on the copper layer. Figure 4 shows the timing sequence of the overall processes.

The proposed system was tested for feature patterning on the 4 inches copper board based on the design shown in Fig. 5 (left). However, the surface of the coated photoresist after the spin coating process was found uneven and the surface roughness was found technically difficult to be approached due to the nature of the photoresist material used (photoresist paint). Despite being diluted in different solutions using distilled water, deionized water and acetone respectively, never the less, due to the viscosity of the photoresist paint, the dilution is difficult to curb the surface roughness and unevenness of the coated layer. The problem persists in the development and etching process, where the coated photoresist layer was still not able to be removed. The complete development of the photoresist layer was not formed even after 10 minutes has collapsed. Therefore, the remainder of the photoresist layer has to be removed by means of manual brushing before etching of the copper layers. Figure 5 shows the copper board after developed (mid) and etched (right).

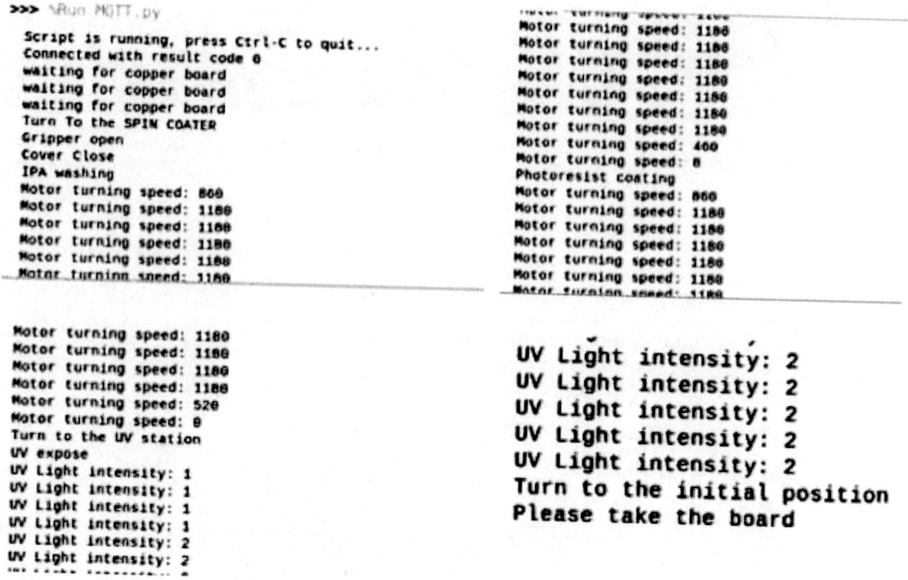

Fig. 6 Visualization of the manufacturing processes on python shell

In spite of the technical difficulty encountered, the microfabrication stations were able to communicate and interact with the remote server hosted on Raspberry Pi. Figure 6 shows visualization of the manufacturing processes displayed on python shell of the remote computer (Raspberry Pi).

4 Conclusions

The work proposed development of wireless interface system for microfabrication stations to a remote server for real-time process visualization. Compared with traditional machine look up, the developed system enables monitoring of all the machines and prompts manufacturer for immediate resolution in real-time. Especially in large scaled production, the mechanism is essential for safeguarding manufacturing down time and ensures the production is able to be delivered on time. The developed system can be easily implemented on any existing stations. Despite of the successful implementation, further experimentation is still required to standardize the application in actual production field. Limitation of the proposed system such as uneven surface thickness due to the use of viscous photoresist solution and in-line process monitoring of the feature size of the devices under fabrication should be included as future improvement of the work.

Acknowledgements The authors wish to thank Ms. Khairun Nisak and Department of Electrical and Electronics Engineering, Tunku Abdul Rahman University College (TARUC) for the support provided to this work.

References

1. Pimpin A, Srituravanich W (2011) Review on micro- and nanolithography techniques and their applications. Eng J 16(1):37–56
2. Huntington MD, Odom TW (2011) A portable, benchtop photolithography system based on a solid-state light source. Small 7(22):3144–3147
3. Tukade TM, Banakar R (2018) Data transfer protocols in IoT—an overview. IJPAM 118 (16):121–138
4. Syafaah L, Minarno AE, Sumadi FDS, Mukti GW (2019) Distance measurement of ESP8266 for control and monitoring in smart home application. J Phys Conf Ser 1381:012058
5. Hamidi EAZ, Effendi MR, Ramdani F (2019) Heart rate monitoring system based on website. J Phys Conf Ser 1402:044003
6. Mohammed MF, Azmi A, Zakaria Z, Tajuddin MFN, Isa ZM, Azmi SA (2018) IoT based monitoring and environment control system for indoor cultivation of oyster mushroom. J Phys Conf Ser 1019:012053